Is The Universe an App?

Schema huius præmissæ diuisionis Sphærarum.

Exploring the Physics of Awareness

Also available from the MSAC Philosophy Group

Spooky Physics

Darwin's DNA

The Magic of Consciousness

The Gnostic Mystery

When Scholars Study the Sacred

Mystics of India

The Unknowing Sage

String Theory

In Search of the Perfect Coke

You Are Probability

The Paranormal Image

Is The Universe An App?

Schema huius præmiſſæ diuiſionis Sphærarum.

Exploring the Physics of Awareness

Mt. San Antonio College
Walnut, California

First Printing: 2014

ISBN 978-1-56543-803-3

MSAC Philosophy Group
Mt. San Antonio College
1100 Walnut, California 91789 USA

Website: *neuralsurfer.com*

Email: *neuralsurfer@yahoo.com*

Authors: *Andrea Diem-Lane and David Christopher Lane*

Type font: *Palatino | point size 11*

The Runnebohm Library Series, in conjunction with the MSAC Philosophy Group, is designed to provide books, magazines, circulars, and pamphlets on science, philosophy, and religion for free via PDF downloads and on various computational applications, including Apple's iBooks, Amazon's Kindle, and Google's Library. Already published textbook titles include *Quantum Physics; Evolution; Neurology; Computers; Mysticism; Radhasoami; Shabd; Skepticism; Sach Khand;* and *Cults.*

Dedication

To our two boys:

Shaun-Michael and Kelly-Joseph

Table of Contents

Acknowledgements

Andrea and I would like to express our deepest thanks to Frank Visser and *Integral World* for publishing a large number of our articles over the years and for encouraging us in exploring the frontiers of neuroscience, quantum physics, and evolutionary biology.

The MSAC Philosophy Group

MSAC Philosophy Group was founded at Mt. San Antonio College in Walnut, California in 1990. It was designed to present a variety of materials--from original books to essays to websites to forums to blogs to social networks to films--on science, religion, and philosophy. In 2008 with the advent of print on demand and cloud computing, the MSAC Philosophy Group decided to embark on an ambitious program of publishing a large series of books and magazines. Today there are well over 100 distinct magazine titles and 50 book titles. In addition, the entire MSAC database is now being put online via Amazon's Kindle, Barnes and Noble's Nook, Google's eBooks, and Apple's iBooks. A special mobile app called Neural Surfer Films is now available for Apple's iPhones and iPads, as well as one for Android operating systems on smart phones and tablets. *The Runnebohm Library* contains works on Einstein, Turing, Russell, Crick and other luminous thinkers. Some of the more popular titles include, *Darwin's DNA: A Brief Introduction to Evolutionary Philosophy* and *Global Positioning Intelligence: The Future of Digital Information*. Over 200 original films touching on quantum theory, neuroscience, and evolutionary biology have also been produced and are free for download via *Neuralsurfer Vision* on *YouTube*. Finally, *The Runnebohm Library* is in the process of producing a number of highly interactive texts that will include embedded video, games, and interactive feedback loops.

Introduction

This book is the first in a new series devoted to exploring the latest developments in neuroscience, quantum theory, and evolutionary biology. Most of the essays here were first published in Europe on Frank Visser's *Integral World*. Our major interest is in understanding how consciousness evolved as a virtual simulator and why it is so important to human cognition and advancement.

While there have been some remarkable developments in evolutionary psychology, a field previously known more controversially as sociobiology, there hasn't been the same attention given to philosophy. Historically, this may be due to the fact that Herbert Spencer, an early champion of fusing philosophy and evolution and a quite popular advocate of such during his lifetime, became something of anathema during the latter part of the 19th and early 20th century because of some of his more controversial views, particularly Social Darwinism. As the entry on him in *Wikipedia* notes: "Posterity has not been kind to Spencer. Soon after his death his philosophical reputation went into a sharp and irrevocable decline. Half a century after his death his work was dismissed as a 'parody of philosophy' and the historian Richard Hofstadter called him the 'the metaphysician of the homemade intellectual and the prophet of the cracker-barrel agnostic.'"

Combining philosophy with evolution can be fraught with peculiar dangers, not the least of which is a tendency towards what Dennett has called "cheap reductionism," explaining away complex phenomena instead of properly understanding it. Nevertheless, it is even more troublesome to ignore Darwinian evolution because it illuminates so many hitherto intractable problems ranging from medicine to ethics.

The new field of evolutionary philosophy, unlike its aborted predecessors of the past, is primarily concerned with understanding why Homo sapiens are philosophical in the first place. It is not focused on advocating some specific future reform, but rather in uncovering why humans are predisposed to ask so many questions which, at least at the present stage, cannot be answered. In other words, if evolution is about living long enough to transmit one's genetic code, how does philosophy help in our global struggle for existence?

To answer that question and others branched with it, one has to deal with the most complex physical structure in the universe—the human brain. Because it is from this wonder tissue, what Patricia Churchland has aptly called "three pounds of glorious meat," that all of human thought, including our deep and ponderous musings, is built upon. Take away the human brain and you take away all of philosophy.

Therefore, if we are to understand why philosophy arose in the first place, we have to begin with delving into the mystery on why consciousness itself arose. And to answer that question we first have to come to grips with Darwin's major contribution to evolutionary theory—natural selection. Why would nature select for awareness, especially the kind of self-conscious awareness endemic to human beings, when survival for almost all species is predicated upon unconscious instincts? What kind of advantages does self-reflective consciousness confer that would allow it to emerge and develop over time?

1 | *Is Consciousness Physical?*

The feeling that consciousness is not physical or that consciousness is distinct from our body is a very common one. Indeed, the very sensation of awareness almost prima facie convinces us that our "I" is not merely an "it." John Archibald Wheeler, the renowned physicist who once mentored Richard Feynman, famously quipped: "The universe and all that it contains ('it') may arise from the myriad yes-no choices of measurement ('bits')." For some who extended Wheeler's line of thinking even further, basing everything upon an informational notion of matter (versus one where matter "forms" data), the ultimate idea was that the universe wasn't at all physical, but rather the condensation of a mathematical reality which in itself was formless, akin in some ways to Plato's notion of universal ideals or archetypes.

In Eastern philosophy, particularly certain strands of Advaita Vedanta, the world is indeed an idea and whatever physicality we attribute its causation is merely an illusion—a maya, something which betrays its real origin. Ramana Maharshi, the famous South Indian sage made famous in Paul Brunton's breezy bestseller, *A Search in Secret India*, is a particularly convincing advocate of this *Consciousness as Reality* position.

On the other side of the equation, a number of neuroscientists and philosophers have becoming increasingly interested in trying to explain consciousness without resorting to any sort of spiritual or metaphysical explanations. Thinkers such as the late Francis Crick, Gerald Edelman, V.S. Ramachandran, Paul and Patricia Churchland, and John Searle have in their differing ways attempted to explain how the brain gives rise to our self awareness. For these thinkers there is no mysterious "ghost" in the machine, but rather something much more straightforward. As Crick clearly stated in his 1994 book, "The Astonishing Hypothesis is that 'You,' your joys and your sorrows, your memories and your ambitions, your sense of personal identity and free will, are in fact no more than the behavior of a vast assembly of nerve cells and their associated molecules. As Lewis Carroll's Alice might have phrased it: 'you're nothing but a pack of neurons.' This hypothesis is so alien to the ideas of most people alive today that it can be truly called astonishing."

Ken Wilber, unlike Crick, has long argued for something *more* than mere physicality to explain consciousness. As he suggests in his 1997 paper, "An Integral Theory of Consciousness":

"This `simultracking' requires a judicious and balanced use of all four validity claims (truth, truthfulness, cultural meaning, functional fit), each of which is redeemed under the warrant of the three strands of valid knowledge acquisition (injunction, apprehension, confirmation) carried out across the dozen or more levels in each of the quadrants—which means, in shorthand fashion, the investigation of sensory experience, mental experience, and spiritual experience: the eye of flesh, the eye of mind, and the eye of contemplation: all-level, all-quadrant. And this means that, where appropriate, researchers will have to engage various injunctions that transform their own consciousness, if they are to be adequate to the postformal data. You cannot vote on the truth of the Pythagorean Theorem if you do not learn geometry (the injunction); likewise, you cannot vote on the truth of Buddha Nature if you do not learn meditation. All valid knowledge has injunction, apprehension, and confirmation; the injunctions are all of the form, `If you want to know this, you must do this'—and thus, when it comes to consciousness studies itself, the utterly obvious but much-resisted conclusion is that certain interior injunctions will have to be followed by researchers themselves. If we do not do this, then we will not know this. We will be the Churchmen refusing Galileo's injunction: look through this telescope and tell me what you see."

Although I have long sided with Wilber's approach since I first started seriously reading him in 1980 when I was attending graduate school, I think Crick's reductionist approach will end up yielding more fruitful results. I say this, even though I readily concede that an all out "take every level, strand, what have you" approach sounds eminently reasonable at first glance. But this "simultracking" AQAL approach a priori assumes positional truths that have yet to be proven. It may be one thing to say "yes, let's meditate and find out what arises," but it is quite another to then invoke reifications such as "Buddha Nature" and say axiomatically, "you cannot vote on the truth of Buddha Nature if you do not learn meditation."

This isn't science. This is theology dressed up in the guise of science.

At this juncture why make such categorical leaps of logic when it could well be that you don't need to learn meditation at all to have such experiences? Even Frits Staal, my old philosophy professor at the University of California, Berkeley, opined in his 1975 book, *Exploring Mysticism*, that it may be possible that other methods besides meditation (like drugs, what he termed the "easy" path) may elicit similar mystical experiences. The point is not whether Staal is ultimately correct or not, but rather that the question of what these internal experiences indicate is still an open query. That mystics from varying traditions have practiced concentrative techniques for centuries doesn't tell us anything whatsoever about their ultimate ontological truth value, just as those having wonderful epiphanies after quaffing large doses of wine doesn't reveal the "truth" about its origination.

This point was personally driven home to me back in the summer of 2004 when I lost my sense of smell. I have been a life-long surfer and because of that my sinuses have suffered. As any devoted surfer will tell you, the waters off Southern California can wreak havoc on your health, especially if you surf (as I have done) after a heavy rain or during red tide. Because of recurring sinusitis I developed enlarged polyps that obstructed my nasal passages causing my sense of smell to completely disappear. I literally couldn't smell a thing, which can be both a blessing and a curse.

What I hadn't realized at the time was how instructive this episode would be in my understanding of how consciousness may have arisen. Smell is a truly amazing sense and when it is fully functioning it provides innumerable forms of information, including deeply resonant emotional evocations. It is like a world unto itself. On occasion, particularly after getting a prednisone shot, I would get my sense of smell back for a week or two. It was exhilarating for me as a hitherto closed off realm was reopened. Now, while my olfactory experiences seemed to border on the numinous when I could smell (Indian food, sea weed, even my boat were heavenly aromas), the fact remains that it was a purely physical event that triggered them.

While I was wafting in varying "sacred" scents, it felt as if my experiences were not at all physical in the mundane sense of the term. The experience seemed to transcend it neuronal and nostril origins. Yet, I know that all such experiences were generated from physiological structures within my own body.

This got me to thinking about consciousness in general and how a cluster of polyps could be so pivotal in determining whether I had access to hidden worlds. This, of course, is analogous to those engaged in contemplative traditions who by dint of their intense meditative practices claim to have access to realms of experience not accessible to others. And who often claim that they have encountered trans-personal or meta-physical states of being. Many further argue that these sojourns are beyond the triune brain's ability to produce them. While I think it is quite true that meditation can elicit all sorts of phenomena that may otherwise remain unknown, it does not follow that what generates such experiences is correctly revealed by the contemplative practice itself. It could be, like one's sense of smell, due to quite physical causes. I realize that when we are having these mystical experiences they do indeed seem to transcend physical causation, but that sensation of ours could just as well be part and parcel of the peculiar biochemistry within our own nervous systems.

A reductionist methodology is actually best suited to determine the physicality of such events. This doesn't mean that mystical encounters are physical, but only that taking an *Occam's Razor* approach forces us to find the necessary mechanisms in the human brain for generating such illuminating displays. Ironically, reductionism is mysticism's most powerful ally. Why? Because reductionism by force of its severe focus on physical causes eliminates non-viable candidates.

Take the so-called miracles of Sathya Sai Baba as one illustrative example of how reductionism, and not merely uncritical phenomenology, can uncover so-called spiritual secrets. Sai Baba followers claim that their Indian guru has the power to produce all sorts of fantastic miracles, including producing vibuti, jewelry, and other objects out of thin air. There are hundreds of thousands of followers who believe that Sai Baba has access to realms beyond the known laws of physics.

The truth, however, behind Sai's miracles is that they are nothing more than sleight of hand magician tricks. We know this now because Sai's trickery has been caught on film. There have also been eyewitness reports of incidents when Sai Baba botched his vibhuti palming, accidentally dropping the tiny tablets which he snapped that contained dried and burnt cow dung. As the *Findings* website explains:

"During darshan, Sai Baba carries vibhuti in tablet form between the third and fourth fingers of his right hand, with spare tablets in the hand holding up his robe. He crushes a tablet when required, and transfers tablets during the taking of letters. I have watched this happen innumerable times. Once on the mandir porch he dropped a tablet in front of me, and told a member of the Trust to 'Eat it Quickly!' For years I had enjoyed the privilege of being called to the interview room and had spent every moment there, focused only on Swami's face; until David [Bailey] suggested that I shift my attention to his hands. Watching rings, watches and other trinkets being palmed, or pulled out from the side of chair cushions, and seeing vibhuti tablets held between fingers before being crushed and 'manifest' was a horrifying revelation, a personal catastrophe for me. I had given up my life, my marriage, husband, children, home, career and homeland because of my love for Sai Baba - only to find trickery at the epicenter of all I held."

Thus, instead of some transcendent phenomenon (needing an appropriate method to apprehend the same) we discover that Sai's miracles are nothing of the sort. But who was it that unearthed this secret? Who among Sai's devotees finally explained what was once considered inexplicable? It was the doubters, the skeptics, the "reductionists"—the very people that many religionists rail against.

As James Randi, Michael Shermer, and other professional skeptics have repeatedly pointed out, the best person to analyze a miracle worker is a magician, not a believer. We are too easily duped by magical tricks, too easily deceived by the confusion of cause and effect, too easily mistaken by the conflation of an object with its image. In other words, the best work in revealing Sai Baba's supposed divine powers wasn't done by taking a spiritual approach, but rather by grounding his claims in the here and now—in the very physicality of the time and place he performed them. When this was done his "paranormal" powers turned to be anything but.

I mention this primarily because Wilber's non-reductionist tendencies don't elicit the kind of groundbreaking information he supposes. Is it any coincidence that Wilber's overly enthusiastic endorsement of the American born spiritual master, Da Free John hindered, versus helped, uncover the guru's nefarious activities with a number of American women? Wilber wasn't on the front lines revealing the ins and outs of Da's very questionable relationships with his disciples. He was literally too busy praising how Da's genius was misunderstood. Is this kind of naivety something we should expect from a pandit who prides himself on being a "critical" thinker?

I confronted Ken Wilber about Da Free John on two occasions in the mid-1980s when we met over dinner in San Francisco and when we met again a couple of years later over breakfast in Del Mar, California, where was I was then living. I happen to genuinely like Ken and I have always found him to be immensely engaging and charming. And even though he listened to my criticisms of Da (to Ken's credit, he did write me once and say that Da was a "fuck-up"), I found him to be surprisingly gullible when it came to appraising spiritual masters. His present association with Andrew Cohen only reconfirms my suspicion about Ken's greenness in this area.

That Cohen's own mother chose to write a scathing expose' of him warrants more than just passing attention.

While I certainly applaud Wilber's efforts to be as all inclusive as possible, his approach doesn't engender much confidence when you look over his track record in dealing with spiritual masters who make extraordinary claims. In fact, I sometimes wince when I re-read some of Wilber's endorsements of Da Free John's writings. Do I really want him to be my critical guide in appraising the supernatural, even if I appreciate his empathetic tolerance that nobody is wrong 100% of the time?

Generally skeptics, not believers, further the cause for the paranormal because they systematically demand evidence that can indeed withstand rational scrutiny. As Carl Sagan rightly stated, "Extraordinary claims demand extraordinary evidence."

And, therefore, the spirit of reductionism can be quite conducive in ferreting out pseudo candidates for the transcendent. More precisely, critics are religions' best friend, provided that such religions wish to know whether their respective truth claims are genuine or not.

Wilber's tendency towards inflationary rhetoric is well documented, but his modeling systems also have a problem of too readily accepting truth claims that may turn out to be simply the product of folk psychology.

For instance, as Wilber writes in "An Integral Theory of Consciousness":

"Subtle energies research has postulated that there exist subtler types of bio-energies beyond the four recognized forces of physics (strong and weak nuclear, electromagnetic, gravitational), and that these subtler energies play an intrinsic role in consciousness and its activity. Known in the traditions by such terms as prana, ki, and chi—and said to be responsible for the effectiveness of acupuncture, to give only one example—these energies are often held to be the `missing link' between intentional mind and physical body. For the Great Chain theorists, both East and West, this bioenergy acts as a two-way conveyor belt, transferring the impact of matter

7

to the mind and imposing the intentionality of the mind on matter."

While it worthwhile to acknowledge that ancient and modern religionists believe in forces such as "prana" or "chi," it does not at all follow that these forces are indeed what the proponents claim they are.

Is there any convincing evidence to suggest that these subtle energies exist "beyond the four recognized forces of physics?" Again, believers are not the best sources for "testing" such claims, nor does it logically follow that you have believe in prana first in order to test it. Rather, a more doubting approach to the subject would force those who make such extraordinary claims to step up and produce their extraordinary pieces of evidence.

As we have seen repeatedly in the psychic world, whenever a so-called master is asked to prove his case it turns out he cannot. Uri Geller, for example, completely failed on national television when he appeared on Johnny Carson's *Tonight Show* and was unable to bend even one spoon.

The late Peter McWilliams, author of the famous book *Life 102: What To Do When Your Guru Sues You*, discovered to his chagrin that his spiritual master, John-Roger Hinkins, didn't have the ability to read students' minds at all, but rather the guru had installed an elaborate taping system throughout his mansion so he could eaves drop on conversations in other rooms.

Perhaps this may explain why some scientists feel that the most promising way to tackle the subject of consciousness is by a process of eliminative materialism. Simply put, if the phenomena cannot be explained fully and comprehensively by mathematics, then one turns to physics, and if that too is incomplete, then to chemistry, then to biology, then to psychology, then to sociology, etc. The old joke is that if none of these academic disciplines can explain it then it is perfectly okay to say, "Well, God did it."

In other words, try to explain it simply first. This is why Occam's Razor (*Entia non sunt multiplicanda praeter necessitatem*, "Don't multiply entities beyond necessity") is such a powerful weapon in science and why ideas such as

Hume's Maxim ("That no testimony is sufficient to establish a miracle, unless the testimony be of such a kind, that its falsehood would be more miraculous than the fact which it endeavors to establish.") and *Laplace's Dictum* ("The weight of evidence for an extraordinary claim must be proportioned to its strangeness.") serve as helpful guide posts.

Edward O. Wilson has captured this same spirit in his book, *Consilience*, which suggests that a unification of the sciences and humanities should be predicated upon a deep and robust understanding of what Bertrand Russell called "natural facts."

It is not that only simple things exist or that there may not be something beyond the rational mind, but only that to genuinely uncover these transcendent phenomena one must eliminate lower level categories first.

When we scientifically advanced in astronomy, medicine, and physics we replaced the old and outdated concepts of our mythic past with new and more accurate terminology that reflected our new found understanding of our body and the universe at large.

Thus, instead of talking about Thor, the Thunder God, we talked instead about electrical-magnetic currents. Thus, instead of talking about spirits as the causes of diseases, we talked about bacteria and viruses. Thus, instead of talking about tiny ghosts circulating throughout our anatomies pulling this or that muscle, we talked about a central nervous system.

In sum, we "eliminated" the gods or spirits in favor of more precise and accurate physiological explanations. Hence, the term: "eliminative materialism."

As a materialistic explanation evolves over time, it will either eliminate or reduce hitherto inexplicable phenomena down from the celestial region to the empirical arena. And in so doing, help us to better understand why certain events transpire in our body, in our mind, in our society, and in our world. *Eliminative materialism is reason writ large.*

The glitch, though, is that we have allowed eliminative materialism to change our thinking about almost everything *except* ourselves.

When it comes to understanding our own motivations, we have (as the Churchlands' point out) resorted more or less to "Folk Psychology," utilizing terms such as "desire," "motivation," "love," "anger," and "free will," to describe what we believe is happening within our own beings.

The problem with that is such terminology arises not from a robust neuroscientific understanding of our anatomies but rather arises from a centuries old mythic/religious comprehension of our very consciousness. And that's the rub.

Where we have moved away from such religious goo speak in the fields of physics, astronomy, chemistry, and biology, in talking about ourselves we are still stuck in pre-rational modes of discourse. Where astronomy reflects the latest theories of the universe, where medicine reflects the latest theories of diseases, in talking about ourselves we tend to reflect ancient theories of human psychology.

And in order to get a better understanding of human consciousness, neurophilosophy argues that we focus our attention on developing a more comprehensive analysis of the brain and how it "creates" self-reflective awareness.

In so doing, we can then come up with a more neurally accurate way of describing what is going on within our own psyches (pun intended). Thus, instead of using the term "soul" we might instead use phase-specific words to describe the current state of awareness that are more neurologically correlated.

We have already done this slightly when it comes to headaches. Due to our increased attention to various pains and to the various drugs that are effective in treating them, we have become more aware of how to differentiate and thereby treat varying types of head pain—from Excedrin (very good for migraines because of the caffeine and aspirin combination) to Advil (very good for body and tooth aches).

Hence, the neurophilosophical way to understand one's "soul" is to ground such ideas in the neural complex.

Now if consciousness cannot be explained sufficiently (Occam's Razor only works if it can indeed explain the given phenomena accurately) with just recourse to the brain, then that form of reductionism has actually helped, not hindered, the case for religionists or transpersonalists since it has exhausted the neuronal possibilities.

But has that happened yet? No.

We are in a similar situation today with consciousness as we were back in the 1920s when it came to unraveling the secret code behind life and inheritance. We might be surprised to learn that back in the early part of the 20th century a large number of thinkers argued that it would be impossible to unlock the secrets of genetics because life was fundamentally based on something not physical, what Bergson had called "élan vital."

It wasn't AQAL inspired scientists who unlocked the double helix structure to deoxyribonucleic acid, but rather two radically reductionist biologists, James Watson and Francis Crick (trained in chemistry and physics), who unraveled its molecular secrets.

And who inspired them to do such? Interestingly, it wasn't a professional biologist at all, but rather the famous quantum physicist Erwin Schrödinger who published a highly influential book entitled *What is Life?* in 1944 based upon his public lectures at Trinity College in Ireland. In just two lines, Schrodinger captured the fundamental approach that he believed would lead to figuring out the problem of life and inheritance. He wrote,

"How can the events in space and time which take place within the spatial boundary of a living organism be accounted for by physics and chemistry?' the preliminary answer which this little book will endeavor to expound and establish can be summarized as follows: the obvious inability of present-day physics and chemistry to account for such events is no reason at all for doubting that they can be accounted for by those scientists."

Schrodinger went on to say essentially that if we take a purely physicalist purview (read: reductionist), then the secrets of the living organism must be somewhere above an atom and below a single cell. In other words, let us not prematurely venture off into metaphysical speculations when the hard physical work still awaits us. It is also no accident that one of Schrodinger's chapters from the same book is entitled, "The Physical Basis of Consciousness."

If we start first with trying to understand how the brain can produce consciousness then even if we do not succeed we will as a byproduct have a much deeper grasp of what it is that the brain does in relation with our awareness.

A simple analogy may be useful here. If you have a particular ailment and go to your local physician, it would be helpful if he/she could eliminate any physical causes for such before unnecessarily branching off into more complicated areas. Because if it turns out to indeed be a broken bone or a nasty bacterial infection, then the doctor hasn't wasted his and your time because he focused on the fundamentals first. Likewise, if cognitive scientists spend their time first on trying to understand how our brains create consciousness then if they do succeed we can come up with a whole series of alternative modeling scenarios for why we think the way we do. If, however, such attempts turn out to be unsuccessful and we have exhausted each and every physical avenue, it will only clarify the difficult task ahead. Again, such a reductionist tendency will actually add to our progressive knowledge of consciousness by eliminating less viable candidates. To use political parlance, we have to "vett" our varying physical theories of consciousness first before we go about hyping higher and more complicated emergent structures.

This bottom up approach isn't anti-informational and it isn't anti-religion. It is, rather, quite practical. Ask yourself next time you take your car into the shop what "approach" you want your mechanic to have on your engine.

Should he really spend his time looking for unseen forces like goblins or gremlins first or as part of his "simultracking" model? I think not. Now, to be fair, there really could be a demon under the hood of your automobile (and we are not talking about inflated gas prices), but in order for that to be the case it might be judicious to make sure it isn't a faulty spark plug first.

If we were to follow Schrodinger's lead, as given to us about organic life, we might be tempted to say that consciousness must be located in something that is less than our body, but larger than a neuron. For some neuroscientists, this can be narrowed down even further to that area known as our brain. Obviously something unique is going on within the confines of that cranial cavity, that three pounds of wonder tissue. But even here we might confront some troubling issues, as Antonio Damasio has indicated in his books, since the brain doesn't exist in isolation but is rather part and parcel of a larger environment, which itself is housed in a larger eco-system—each of which act as moment to moment feedback loops.

Partly because of this difficulty, a growing segment of the cognitive science community has partitioned the problem into smaller chunk size problems. The well known neurologist from UCSD, V.S. Ramachandran, for instance, has focused primarily on the visual cortex, suggesting that if we could understand this one component of consciousness we might have an easier time scaling up to tackle the whole problem. This makes eminently good sense, since even if understanding vision doesn't unlock the secrets of awareness, it does provide very useful knowledge about that most vital part of our senses. As Christof Koch explains in a co-authored paper with Francis Crick entitled, "Consciousness and Neuroscience":

"How can one approach consciousness in a scientific manner? Consciousness takes many forms, but for an initial scientific attack it usually pays to concentrate on the form that appears easiest to study. We chose visual consciousness rather than other forms, because humans are very visual animals and our visual percepts are especially vivid and rich in information. In addition, the visual input is often highly structured yet easy to control. The visual system has another advantage. There are many experiments that, for ethical reasons, cannot be done on humans but can be done on animals. Fortunately, the visual system of primates appears fairly similar to our own (Tootell et al., 1996), and many experiments on vision have already been done on animals such as the macaque monkey. This choice of the visual system is a personal one. Other neuroscientists might prefer one of the other sensory systems."

If you think about your awareness you will begin to realize that the notion of its unity may in itself be illusory. Combinational systems, when aligned properly, can indeed give the impression of unification. A rudimentary example that we might all be familiar with is our home entertainment centers, whether it is an old speaker system connected to a C.D. player or a newer surround sound system fully integrated with your new high definition flat screen television, replete with a Blu-ray DVD player. You can tinker with your sound system so that although each speaker is in different corners in your room the music will feel like it is coming integrally from the center.

To buttress this idea further I often given an analogy in my philosophy classes that came to me from watching varying bands play at the Tomorrowland Terrace in Disneyland. Imagine that there are different musicians, each playing different instruments, tuning up. Finally, after much tuning up each musician starts to play in harmony with the other players until finally they unite into one beautiful song. Further imagine that once this unification occurs a conductor appears waving his baton acting like he has been conducting the music all along. However, you know that he only arose after all the musicians were in harmony. The conductor wasn't the cause of the music. He was, most pointedly, the result of the music, even if he or she acts as if they were the maestros all along.

If consciousness is indeed the result of a combinational connection of varying parts of the body system, but most particularly the brain, then it would make great sense to try to understand each of those distinct, but not segregated components first.

Or, to put it more bluntly: If our seeing is physiologically based and our hearing is physiologically based and our smelling is physiologically based and our tasting is physiologically based and our feeling is physiologically based, then is it really that much of a stretch to think that our "being" is physiologically based as well?

I don't think so, but we will never find this out unless we can duly eliminate all those physical causes first. In other words, if one truly thinks that consciousness is the result of something beyond the brain/body complex, then it is vitally important to make absolutely certain that awareness cannot be merely the outcome of physiological processes.

In sum, we have to fully explore physicality before we can invoke higher emergent structures. If we fail to do so, then we run the very probable risk of being just as duped as those who bought into Sai Baba's miracles because they too readily accepted something as mystical when, in point of fact, they were merely bad parlor tricks.

I remember one of my former philosophy students once telling me that my skepticism was misguided since he could leave his body at night during sleep and travel to other parts of the globe and remember many key details from his nocturnal sojourns. I told him of my own similar excursions, but that I (like Blackmore, Feynman, Faqir Chand, and many others) felt that such experiences were not objectively real but were rather the projections internally of our own mind.

But, since I could easily be wrong (skepticism may be many things, but closed minded should not be one of them), I then proposed a little experiment to him that I had cribbed from my readings in parapsychology. I said that I would put a seven-digit number on my office wall and that over the next few nights he could astrally travel and jot the numbers down to his memory. Now this wasn't a fail proof test, but we both thought that if he succeeded it would certainly warrant further attention.

However, to his disappointment, every time he went to my office in his nightly O.B.E.'s he ended up securing a completely wrong number. He wasn't even close. Now this doesn't prove that O.B.E.'s are merely brain hallucinations, but it does throw the burden of proof back on where it belongs. If you make a claim then, like that short-lived reality television show on MTV, you have to back it up. Otherwise, we are back to square one and speculation 101.

While there have been many reports in the literature about how there have been some remarkable breakthroughs in parapsychology (such as Dean Radin's book, *The Conscious*

Universe), a closer scrutiny of the literature has shown a field which has offered up terribly dismal results.

This doesn't mean, however, that we should give up the search for such things, but only that the most expedient way to progress in this field is to be as critical and skeptical as possible of any extraordinary claim. We too often succumb to what Paul Kurtz called the *Transcendental Temptation*. In our overly eager desire to believe in the mystical, we tend to accept less than convincing evidence since it buttresses our already cherished hopes and desires. It is naturally difficult to contravene what we wish to be the case.

I remember one telling incident that occurred in my own family. An elder relation named John chided me about my tendency to doubt claims of purported telepathy. He was convinced that such powers existed and knew from his own experience that I was wrong to be so questioning. John proceeded to tell me of when he went to see a psychic at a local gathering and the psychic told him the exact serial numbers off his one dollar bill which was buried in his wallet which, he emphasized, was deep within his front pocket. John then provokingly exclaimed, "Explain that Mr. Skeptic."

I thought for a second and then chimed back, "How long do I have to solve it?"

John, looking a bit perplexed said, "How much time do you need?"

"About thirty seconds or so," I responded.

"What?" said John.

I then asked a very simple question to John, "Did you pay any money to see this psychic showing."

John replied, "Why, yes."

I then said, "Did they give you back any change"?

John now looking quite agitated replied, "Well, yep."

I then went in for the jugular, "Was the change in the form of a dollar bill or bills?"

At this point, John knew that I knew that he had been had. John was duped, but even though he was a pawn in a very old magic trick, he still held out that surely some psychics knew things beyond our five senses, even if they occasionally succumbed to less honest gimmicks.

Perhaps the most recalcitrant issue in consciousness is that it provides us with a keen sense of dissociation and therefore its very intangibility appears resistant to a merely physical explanation. Awareness doesn't feel "bodily," except of course on those occasions when we have a toothache, a headache, or any other "associative" ailment tied with our body. How can something so ethereal be the result of something so material?

But this is precisely why our own experiences should not be the sole criterion for appraising how consciousness arises. My experience every morning tells me that the sun rises at dawn but as we now know my inference about the sun is mistaken. The sun doesn't "rise". Rather, our planet earth rotates and this rotation is what causes the impression of a "rising" sun. Yes, my experience may tell me otherwise, but in order for me to have a richer and larger understanding of how the sun and the earth actually operate I have to go beyond my own limited experiences and appeal to measurements that contravene what I think I already know.

Likewise, whatever certainty have that consciousness must be non-physical has to be tempered with the very simple rejoinder that my experience may indeed be both illusory and wrong. The latest studies in neuroscience have come up with some startling insights into how easily we deceive ourselves when it comes to visual perceptions. Neuroscience has also come up with some startling conclusions about why we have self-awareness and how mirror neurons may have evolved to provide us with a dual function. As V.S. Ramachandran explains in a now famous article, "The Neurology of Self Awareness:"

"The discovery of mirror neurons was made by G. Rizzolati, V. Gallase and I. Iaccoboni while recording from the brains of monkeys performed certain goal-directed voluntary actions. For instance when the monkey reached for a peanut a certain neuron in its pre motor cortex (in the frontal lobes) would fire. Another neuron would fire when the monkey pushed a button, a third neuron when he pulled a lever. The existence of such Command neurons that control voluntary movements has been known for decades. Amazingly, a subset of these neurons had an additional peculiar property. The neuron fired not only (say) when the monkey reached for a peanut but also when it watched another monkey reach for a peanut!"

"These were dubbed 'mirror neurons' or 'monkey-see-monkey-do' neurons. This was an extraordinary observation because it implies that the neuron (or more accurately, the network which it is part of) was not only generating a highly specific command ('reach for the nut') but was capable of adopting another monkey's point of view. It was doing a sort of internal virtual reality simulation of the other monkeys action in order to figure out what he was 'up to'. It was, in short, a 'mind-reading' neuron."

"Neurons in the anterior cingulate will respond to the patient being poked with a needle; they are often referred to as sensory pain neurons. Remarkably, researchers at the University of Toronto have found that some of them will fire equally strongly when the patient watches someone else is poked. I call these 'empathy neurons' or 'Dalai Lama neurons' for they are, dissolving the barrier between self and others. Notice that in saying this one isn't being metaphorical; the neuron in question simply doesn't know the difference between it and others."

"Primates (including humans) are highly social creatures and knowing what someone is "up to"—creating an internal simulation of his/her mind—is crucial for survival, earning us the title 'the Machiavellian primate'. In an essay for Edge (2001) entitled 'Mirror Neurons and the Great Leap Forward' I suggested that in addition to providing a neural substrate for figuring out another person's intentions (as noted by Rizzolati's group) the emergence and subsequent sophistication of mirror neurons in hominids may have played a crucial role in many quintessentially human abilities such as empathy, learning through imitation (rather than trial and error), and the rapid transmission of what we call 'culture'. (And the "great leap forward"—the rapid Lamarckian transmission of 'accidental' one-of-a kind inventions.)"

"I turn now to the main concern of this essay—the nature of self. When you think of your own self, what comes into mind? You have sense of "introspecting" on your own thoughts and feelings and of 'watching' yourself going about your business—as if you were looking at yourself from another person's vantage point. How does this happen?"

"Evolution often takes advantage of pre-existing structures to evolve completely novel abilities. I suggest that once the ability to engage in cross modal abstraction emerged—e.g. between visual 'vertical' on the retina and photoreceptive 'vertical' signaled by muscles (for grasping trees) it set the stage for the emergence of

18

mirror neurons in hominids. Mirror neurons are also abundant in the inferior parietal lobule—a structure that underwent an accelerated expansion in the great apes and, later, in humans. As the brain evolved further the lobule split into two gyri—the supramarginal gyrus that allowed you to 'reflect' on your own anticipated actions and the angular gyrus that allowed you to "reflect" on your body (on the right) and perhaps on other more social and linguistic aspects of your self (left hemisphere) I have argued elsewhere that mirror neurons are fundamentally performing a kind of abstraction across activity in visual maps and motor maps. This in turn may have paved the way for more conceptual types of abstraction; such as metaphor ('get a grip on yourself')."

"How does all this lead to self awareness? I suggest that self-awareness is simply using mirror neurons for 'looking at myself as if someone else is look at me' (the word 'me' encompassing some of my brain processes, as well). The mirror neuron mechanism—the same algorithm—that originally evolved to help you adopt another's point of view was turned inward to look at your own self. This, in essence, is the basis of things like 'introspection'. It may not be coincidental that we use phrases like 'self conscious' when you really mean that you are conscious of others being conscious of you. Or say 'I am reflecting' when you mean you are aware of yourself thinking. In other words the ability to turn inward to introspect or reflect may be a sort of metaphorical extension of the mirror neurons ability to read others minds. It is often tacitly assumed that the uniquely human ability to construct a 'theory of other minds' or 'TOM' (seeing the world from the others point of view; 'mind reading', figuring out what someone is up to, etc.) must come after an already pre- existing sense of self. I am arguing that the exact opposite is true; the TOM evolved first in response to social needs and then later, as an unexpected bonus, came the ability to introspect on your own thoughts and intentions."

Even the most profound spiritual experiences may themselves be the result of brain processes of which we remain unaware. This doesn't discount the beauty or bliss of such numinous journeys, since there are many things we enjoy that are indeed the result of physical machinations. For instance, my fondness for surfing (even with my lack of smell) has not disappeared because I know something about the physics of waves. The majestic beauty of a rose isn't lessened by our deeper grasp of its molecular parts. As Feynman once illustrated when he pointed out to his artist companion that a physicist's understanding of a flower doesn't detract from its beauty, but only adds to it since he can appreciate so many other levels that usually go undetected.

In light of how reductionism actually works when applied to real life situations, I am surprised that there are not more strong advocates of it coming from those most deeply interested in mysticism. Blaise Pascal once wrote that those with little faith will have little doubt and those with great faith with have great doubt. While I appreciate his religious syllogism, I don't think he extends if far enough. The logical consequence of his couplet should end with "And those with infinite faith, will have infinite doubt."

Because it is through doubt and skepticism where more, not less, evidence for the transcendent will arise since such critical scrutiny raises the bar for acceptable proof much higher than those who tend to believe on anecdotes alone.

I find it curious that we resist the physical causation of consciousness, knowing as we do nightly that very tiny chemicals, such as adenosine within our brain, can all too easily make our lucid awareness fall almost imperceptibly to mush. The biochemical basis of our awareness is so evident that even this very article you are reading can prove it. If I multiplied this piece tenfold and you were forced to read each and every word, I am quite confident that within fifty more pages the adenosine would kick in and you would find your luminous waking awareness completely upended and your breathing getting louder with peaceful sounds known to us as "snoring." Of course, for some readers this may have already occurred after page five.

It is one of philosophy's great ironies that if you think long enough about how consciousness is not physical, sleep will eventually take over and resolve the argument for you without any words whatsoever.

And if you find yourself, like Ken Wilber, having the ability to remain "aware" in a non-dual state even while sleeping, then you may want to ask why alcohol can so rapidly screw up such luminosity, as Wilber himself acknowledged in his book *One Taste*. Then again, as I humorously tell my Religion and Science classes at CSULB, the ultimate acid test for transcending the notion that consciousness is purely physical is what is lovingly called the hammer test which has many variations, but which always ends up with the same answer. Why is it that when I hit my head with a hammer it invariably alters my state of awareness, especially if my consciousness is not physical?

Or, as the script to the short film *A Glorious Piece of Meat* summarizes the consciousness paradox:

"I know that my consciousness is more than the sum of my neurons firing; or, at least I think so while my neurons are firing."

2 | The Physics of Being Aware

In response to a recent article I wrote with my wife, Andrea, ("Is Consciousness Physical") Frank Visser was kind enough to give us an initial rejoinder with some pregnant questions. I particularly liked what Visser wrote since it focuses some of the key questions confronting neuroscientists and philosophers concerned with explaining how and why human awareness has evolved.

The first question we have to tackle here is Ken Wilber's position that even if one does indeed discover a neurological correlate to a mystical experience, nay any human experience, it does not capture the subjective nature of what we are experiencing at the time. For instance, it may be true that a dentist can objectively know each and every detail about why your impacted wisdom teeth are causing you so much pain, but his/her narrative, no matter how sympathetic it may sound or read, doesn't allow him/her to get directly into your experience of suffering at that time. As the well-known philosopher, John Searle, at U.C. Berkeley, might put it: a third person description shouldn't be conflated with a first person narrative. The latter, while objective, doesn't reveal the subjectivity or the "interior" nature of the patient's dental pain.

David Chalmers in his widely read book, *The Conscious Mind*, in analyzing this conundrum calls "qualia" the hard problem in the study of consciousness. Chalmers believes the problem is so hard that he has tried to re-introduce a respectable version of dualism in order to resolve the paradox. Owen Flanagan from Rutgers University has suggested that the peculiar nature of consciousness is such that we will never be able to solve its essential mystery.

Other philosophers, however, especially those grounded in the neurosciences aren't so pessimistic and have envisioned a series of pathways by which both the subjective and objective nature of awareness can indeed be explained purely physically. As Patricia Churchland, author of the 1986 breakthrough text, *Neurophilosophy*, explains in her seminal paper, "What Should We Expect from a Theory of Consciousness":

"This question is most pointedly raised in the context of the inverted spectrum problem, and I shall address it in that form. To illustrate, consider the possibility that your color experiences (color qualia) might be systematically inverted relative to mine; e. g. where you see red, I see green, and so on. Noting that there could be systematic behavioral compensation that would cover experiential differences, skeptics have urged that even looking inside would be unavailing. Allegedly, no conceivable test could ever reveal similarity or inversion in our color experiences. The lesson we are invited to draw is that consciousness is intractable scientifically because inter-subjective comparisons are impossible. Some philosophers think that this is not merely a problem about what we can and cannot know, but evidence that consciousness is a metaphysically different kind of thing from brain activity. In addressing this issue, I shall make one assumption: that conscious experiences are in some systematic causal connection to neuronal activity. That is, they are not utterly independent of the causal events in the brain."

"To deny the assumption is to slide into a version of dualism known as "psycho-physical parallelism", meaning that mental events and physical events are completely independent of each other causally, and just happen, amazingly enough, to run in parallel "streams". Normal human color vision is known to depend three cone types, each of which is tuned to respond to light of particular wavelengths. Cone inputs are coded by color-opponent cells in the retina and lateral geniculate nucleus, and project to double-opponent cells in the parvocellular-blob pathways of the cortex. Cells in cortical area V4 appear to code for color regardless of wavelength composition of the light from the stimulus, and appear to subserve the perceptual phenomenon known as color constancy. Lesions to the cortical area known as V4 result in achromatopsia (loss of all color perception), in humans and monkeys. Also relevant to behavioral determinations of differences in experience is the fact that the relations between hue, chroma (saturation) and value (lightness) define an asymmetric solid (Munsell color quality space). This implies that radical differences in perception should be detectable behaviorally, given suitable tests. That is, "A is more similar to B than to C" relations between colors are defined over the Munsell quality space, and if there is red/green inversion, the similarity relations to yellow, orange, pink etc. will not remain the same."

"Given the progress to date, it seems likely that the basic neurobiological story for color processing can be unraveled. For similar reasons, it looks likely that the basic story for touch discrimination and its mechanisms in the somatosensory thalamus and cortex can also be unraveled. For example, it seems evident that if someone lacks "green" cones, or lacks "red/green opponent cells, or lacks a V4, they will not experience the visual perception of green. Comparable circuitry and comparable behavioral discriminations seem to be presumptive of comparable experiences; that is, comparable qualia. (Clark 1993) The skeptic, however, insists that no data—not behavior, not anatomy, not physiology—could ever reveal qualia inversion. Incidentally, what is at issue here is not that minor differences in such things as hue or brightness might go undetected, but that even huge differences, such as red/green inversions or brightness/dullness inversion, might be in principle absolutely undetectable. To approach this matter somewhat indirectly, let us first consider a vivid example where it is evident that a perceptual inversion is likely detectable, namely, inverted "shape" qualia. Could Alphie have "straight qualia" whenever Betty has "curved qualia" (and vice versa) and the difference be absolutely undetectable? In this example, because two modalities—vision and touch—can access the external property, it seems easier to agree that together, behavioral data and wiring data permit us to make a reasonable determination of similarity and differences. That is, the problem is not essentially less tractable than determining whether two people digest food in the same way or whether two cats in free fall right themselves in the same way."

"Much the same is usually conceded for pleasure/pain inversion, and within vision, of near/far inversion in stereopsis. Random dot streograms are already a very reliable determinant of (a) whether a subject has stereopsis at all, and (b) whether there is an inversion between two subjects. If we factor in data from "near/far" cells in V1, then insisting upon absolute undetectability begins to look unreasonable. It seems a bit like insisting that it is absolutely undetectable whether the universe was created five minutes ago, complete with all its geological records, its fossil records, history books, and my memories etc."

"Skepticism carried to that extreme just ceases to be scientifically inter[e]sting, and becomes philosophical in the pejorative sense of the term. In the case of "shape inversion", the skeptic can remain a skeptic by going one of two ways, neither of which helps his sweeping anti-reductionist defense: (1) no qualia—not shape, not temperature, not pain not any qualia-- can be compared across subjects, even to a degree of probability. They are one and all, absolutely incomparable. For all I could ever know, you might experience the color red when I experience pain. (2) The neurobiology of shape qualia (rough/smooth, etc.) can be compared, and perhaps even the neurobiology of stereopsis, but color vision is different. The first appears to depend only on an anti-reductionist resolve, without any independent argument. In that case, we really are looking at a circular argument. The only escape from the circle is to fall into the embrace of dualism—and worse, of the deeply implausible psycho-physical parallelism rendition of that already implausible doctrine. The second makes a major concession so far as qualia in general are concerned. Having conceded that some qualia are scientifically approachable, the skeptic no longer shields subjectivity as such, but only subjectivity for certain classes of experience, namely color vision. This looks far less powerful that the original position, and it starts the skeptic down a slippery slope. For having made this concession, it now becomes easier for the reductionist to push hard on the point that comparisons in receptor properties, wiring properties, connectivity to motor control, and so forth, will augment—as it already does—behavioral data, and allow us to compare capacities across individuals. And similar arguments can be made for other single modality experiences: stereopsis, sound, pain, temperature, feeling nauseous, feeling dizzy etc. That is, as long as awareness of color has a causal structure in the brain— as long as it is not a property of soul-stuff utterly detached from all causal interaction with the brain—data from psychology (e. g. the color-hue relationships) and neuroscience (tuning curves for the three cone types in the retina, wiring from the retina to cortex and intracortically) predicts that big differences in color perception will correlate with big differences in wiring and in neuronal activity. In the context of a more detailed knowledge of the brain in general, rough comparisons between individuals ought to be achievable, subject to the usual qualifications unavoidable in any science."

Wilber tends to see distinct, even if connected, realms between the body, the mind, and the spirit. Indeed, as he first suggested in *Eye to Eye*, there is a tripartite methodological schema which gives rises to differing truth claims, "which means, in shorthand fashion, the investigation of sensory experience, mental experience, and spiritual experience: the eye of flesh, the eye of mind, and the eye of contemplation: all-level, all-quadrant."

Now the problem that naturally arises here is that the eye of flesh and the eye of mind and the eye of contemplation are, from a physicalist perspective merely descriptions of varying functions within one's own single (even if messy and conjoined) brain. The danger here is that Wilber tends to conflate descriptive markers or activities with a hierarchical notion of truth.

Yes, I can focus my attention on the visual field in front of me, such as the computer screen I am looking at, or I can instead divert my gaze and ruminate about mathematical symbols that I can imagine in my head without recourse to anything presently in my visual field. Or, I can close my eyes in order to stop or alter what images seep into my cranium. Or, I can try to shut down my chattering thoughts by focusing on one image or no image at all, or repeating a mantra, or finding the source from which my awareness arises. But underlying all three of these activities isn't a hierarchy of distinctive modalities for truth ascertainment. What we are witnessing, quite literally, is the various ways we use our attention.

Wilber's descriptions run the danger of being taken as reifications (literally making a "thingy" out of a descriptive abstraction). There isn't a "physical" truth versus a "mental" truth existing on separate planes of existence. Likewise, calling something a spiritual truth implies some kind of transcendent ontology, when such is not at all the case. When I meditate I am not necessarily ascending to something trans-physical or trans-mental. Rather, one could just as easily argue that I am merely invoking a different aspect of my own awareness, such as turning my attention to attention itself or relaxing the grasping of objects in my visual field to be the field itself. But all of this can rightly be seen as various aspects of one physical being.

While meditative and contemplative disciplines certainly have much to teach us about how our brains work, I don't think using outdated and outmoded theological conceptual worldviews is helpful in understanding consciousness.

Rather, it seems to be a huge impediment. For example, demon possession is no longer a viable descriptive marker to explain aberrant states of mind. We realize now that such misleading labeling obscured a deeper understanding of how the brain operates under certain psycho-social conditions. Because of our improved neuroscience, professional psychologists have more or less "eliminated" demonic possession as an accurate behavioral term. This doesn't mean that the belief in such has disappeared, but only that rational scrutiny of the phenomenon has demonstrated that this mental state and others like it were misleadingly identified and explained. We know better now because have a deeper and richer understanding through science of how our nervous systems react in varying conditions.

Likewise, terms such as Samadhi, enlightenment, Buddha Nature, while indicative of how certain religious systems described their internal practices, may also be "eliminated" and replaced with more neurologically precise terminology. To keep using outdated and theologically loaded terminology, as Wilber has tended to do, can unwittingly create fake or misleading truth hierarchies where none may exist.

Wilber, to his partial credit, has already admitted to some of this and thus he speaks of his own evolution with his slightly narcissistic appellations of Wilber I, Wilber II, Wilber III, and so on.

But I don't think he has gone far enough. Indeed, he hasn't really come to grips with how his whole AQAL system can be turned upside down and changed quite easily into Edward O. Wilson's reductive *Consilience*, where the exploration of higher realms of existence really does take on a scientific posture and which grounds itself in physics and chemistry and biology first. Wilber is prematurely pontificating about that which is trans-rational, when on closer inspection much of it may simply be sophisticated, if mostly unexplored, machinations of our own cranial capacities.

Do we feel this sense of exploration from Wilber's AQAL maps? No, we get rather the conceited sense that the cartography has already been charted and figured out and all we have to do is fill in the gaps. Wilber's Integral is theology dressed up in scientific jargon. But science is progressive only to the degree that it can be falsifiable.

So, the best thing for Wilber's whole "ism" (pun intended) is its rejection, not its acceptance. In other words, the focus should be on seeing where and when his pontifications are mistaken. His AQAL quilt is only as good as its patches, and if a certain patch of his research is weak or insufficient then that should be immediately shored up. Wilber is pushing so hard for his mosaic that he either forgets or ignores the numerable pieces that don't fit or are severely damaged.

This is why Wilber's reaction to critics and criticism is so telling. Wilber has already been rightly raked over the coals for his take on evolutionary biology, but what is even more worrisome is his reaction to Frank Visser and his website. Just re-read Wilber's varying rants about who should qualify as a critic and you immediately realize that much of his Integral has absolutely nothing to do with science, but everything to do with building up a cultic worldview that doesn't t even withstand his own critique of such things from his earlier book, *Spiritual Choices*.

Yet, the most troublesome aspect of Wilber's theology (and, yes, it is more theology than science) is, ironically, how cheaply reductionist it can be when it suits his lexical needs. His entire take on meme theory is illustrative of this. Labeling varying individuals and their thinking with color codes as a type of short-handed stereotyping isn't insightful but actually borders on a new form of intellectual racism. Instead of calling somebody white trash, you can instead slime him with the "oh, he's at the green meme level." This kind of label reductionism is completely at odds with anything pretending to be "integral." Moreover, it is intellectually lazy, not to mention highly confusing as well, especially in light of how memetic theory was developed by Richard Dawkins and later by Susan Blackmore.

As for whether science can indeed decide the physicality of consciousness I think the answer is affirmative, even despite science's methodological naturalism. Quite simply, if consciousness is indeed beyond physics or anything within its known laws, then no matter how hard we try to ground mind to its neural structures there will always be something missing in such reductions. And, interestingly, this gap will loom even larger because our physical science will be unable to adequately explain it. Thus, one could argue that such a physicalist approach will shine a much more illuminating light upon the problem by showing exactly where, when, and how awareness is not the result of physical properties. But if we forego this grounded scientific quest prematurely because of already accepted quadrant categorizations (the type that Wilber is fond of proposing, even if he talks of some ultimate spiritual union) then we can and will succumb too easily and too readily and too naively to the *Transcendental Temptation*. Or, to invoke Wilber's own parlance, you cannot make a pre-trans fallacy distinction unless you have a deep and rich and nuanced understanding of all that is indeed pre. How else can one determine that which is truly trans?

Or, as I once wrote in my critique of Jack Hislop's highly readable but highly questionable book, *My Baba and I*, substantiating Sathya Sai Baba's supposed miracles, why does the Swami produce trinkets out of his hand and not the world's largest diamond? Or, why not pull a Toyota Camry out of his ear? Because if he did either of those two things it would force all us of take his claims much more seriously, even if we tried our utmost to come up with a common sense explanation.

Science, in other words, can indeed point to that which is not physical because of its ultra focused aim. Science can upend itself quite easily. The fact that it hasn't yet is why we remain so confident in its methods and its discoveries. But if in the future it comes across something that cannot be reduced to the four forces of the universe, we will be forced to reconsider.

But what has happened in the past and what is still happening today is that we want to invoke transcendent explanations too quickly in order to salvage a sense of the numinous, forgetting in the process that even if all things are indeed material bits the mystery of all this (and here comes the pun) isn't lessened by one bit.

The problem isn't with "all is matter" but with our outdated notion of what matter truly is. Matter, as anyone slightly conversant with quantum theory surely know, isn't flat, isn't grey, and it certainly isn't one-dimensional. Matter is as magical and as mystical as any "heaven" described in religious tomes.

To underline our Wittgenstein-like difficulties, imagine that you are a devout Sikh and that after you died you entered into that most exalted of spiritual regions, *Sach Khand* (literally, "the truth realm"). Further imagine that you as soul/spirit were involved in a heavenly conversation with one of your numerous companions. You query, "Wow, we are in the land of truth and bliss. But what are we made of here?"

Your friend responds, "Ah, we are beings of light."

Now for most of us hearing this conversation here on terra firma it sounds uplifting and we might even wish we could have something similar.

But is it really any different right now? What is matter anyways?—from organisms to cells to proteins to molecules to atoms to electrons to light? The most famous equation in modern physics is Einstein's $E=MC2$ which if we pause for a second is as mysterious as anything written in our ancient religious scriptures and measurably more radical.

My point is that the resistance we have to reductionists who say, we are "just matter" is because we tend to think of matter as flat. It is, of course, anything but. Thus maybe the reason we opt for dualism or the idea that something must be "beyond" matter is because we haven't come to grips to the amazing plasticity and mystery inherent in matter itself.

In other words, we are using an extremely outdated and misleading definition of matter and in so doing losing sight of the wonderfulness of what physics and neuroscience is saying. We are not lessened because we are just matter. Just as we wouldn't be lessened in *Sach Khand* if we were made of just "light."

Frank Visser rightly capture our semantic confusion when he queries, "How on earth could cells and molecules lead to felt states? Forget about psi – psychology itself is as paranormal as you can get!"

At first glance it does seem to absolutely amazing how we could get from molecules to self awareness. But, the same could be said about life itself. How can it be that a three-letter sequence of DNA strung together in varying sequences can produce a giraffe, a shark, and a human being? Indeed, rearrange atoms and you can get a chalk board, a cruise ship, an orange, and the moon. But get it we do.

Likewise, getting from a cell to a self aware human being isn't a stretch if we understand how the complex arrangement of atoms can indeed produce things that we cannot possibly imagine to exist, but which played out over time do indeed exist.

Therefore, the problem that we have with the physicality of consciousness is the same resistance that our ancestors had with understanding probabilities and how very simple algorithmic sequences can produce truly astonishingly complex varieties, the likes of which boggle our imagination.

That we cannot imagine how matter produces consciousness tells us how limited our imaginations are when it comes to the wonders of physics. We shouldn't confuse our intuitive limitations with how the world works.

Physics is the most mystical endeavor known to humankind, if one truly comes to grip with the multi-dimensional aspects inherent in any particle that arises.

If one takes thinks of hydrogen and oxygen in isolation, it would be inconceivable to imagine that their combination would bring forth water. But that is precisely what occurs and nothing "more" is needed. Therefore, the "inconceivability" of something shouldn't be used ad hoc as a precursor for invoking the divine. Patience, in other words, is a highly necessary virtue if we wish to avoid making pre/trans leaps.

We have an almost built-in dualism within our awareness which gives us the convincing sense that our selves are not our bodies. This is what Visser is underlining when he mentions that destroying a television set wouldn't destroy television programs. T.V. shows come through the set but are not of it.

However, one could just as easily argue that if consciousness is akin to an electromagnetic wave then stopping production at its source would indeed cancel the television show.

The brain, in this purview, is the production facility and because of its centrality to self-reflective awareness, it seems fairly obvious that if you destroy the central nervous system you have killed consciousness.

Saying consciousness is physical doesn't detract from its majesty in the least, since as we have repeatedly mentioned matter itself is to use Rudolph Otto's religious terminology, *mysterium tremendum* and *ganz andere.*

We have to turnaround our understanding of matter and also how we view ourselves in the process. Matter is multi-dimensional and if the reconfigurations of tiny atoms can give us nature's wide diversity (from a rose to an airplane to a sunset to a cup of java to repeated episodes of *I Love Lucy*), then a complex set of billions of neurons may also give rise to varying degrees of awareness.

Or, we could use reverse engineering to give us a clue about why awareness is directly connected to physics. The difference between a rock and a chimp isn't something transcendent, but rather due to the complexity of atoms and molecules clustered within the central nervous system of our simian friends. Look to the complexity of matter first and you will readily see why and how awareness arises in some material objects and not in others. Invoking gods or spirits or Eros is literally nonsensical, particularly when the physics of neuroscience is still in its infancy.

Before we succumb to the *transcendental temptation*, maybe it would be prudent to show some patience and let out empirical sciences have a deeper stab at the problem first.

3 | Is My iPhone Conscious?

Someone who dreaming says "I am dreaming," even if he speaks audibly in doing so, is no more right than if he said in his dream "it is raining," while it was in fact raining. Even if his dream were actually connected with the noise of the rain.

—Ludwig Wittgenstein

Our conviction that something is real or certain doesn't mean that we cannot be mistaken. How can one convince a skeptic that there really are higher states of consciousness beyond the rational mind? Or, framed in a different way, can science accurately explain the numinous experiences of mystics?

In several previous articles and books (see *Exposing Cults: When the Skeptical Confronts the Mystical*), I have argued that science is indeed sufficient to the task of explaining the paranormal or transcendent. However, I don't think it is very intellectually challenging or interesting to merely repeat one's Wilsonian leanings or intertheoretic reductionisms over and over again.

So, in honor of both Andy Smith and Elliot Benjamin (who have inspired me with their open-mindedness), I thought it might be fun and hopefully useful if I did a complete U-turn and attempted to argue on their behalf. This doesn't mean, of course, that I have somehow radically changed my mind over night and have thrown Edelman, Crick, and Ramachandran under the proverbial bus, but only that if we are to take the scientific study of mysticism seriously we should on occasion argue against our own purview and interests.

Ideally, for instance, one would love to see Richard Dawkins argue against his *God Delusion* hypothesis and side (even if temporarily) with the very theists he lambasts. Or, to more properly contextualize this in light of the *Integral World* website, it would be refreshing to see Ken Wilber assume a purely Darwinian position and self-critique his longstanding position on spirituality. In other words, perhaps we can learn more about any given subject if we allowed ourselves the freedom of going against our own vested ideologies or stratagems.

I often say on the first day of my Critical Thinking courses at Mt. San Antonio, the ultimate goal of this class is to develop the ability to think of good counter arguments against one's cherished position. As F. Scott Fitzgerald essayed in *The Crack-Up*, "The test of a first-rate intelligence is the ability to hold two opposed ideas in the mind at the same time, and still retain the ability to function."

This also brings to mind the famous syllogism, "little doubt, little faith; great doubt, great faith." Which I feel needs to be extended with its logical conclusion: infinite doubt, infinite faith.

We are not going to proceed very far in the study of mysticism (pro or con) if we are unwilling to switch hats from time to time. But before I do that, I want to first "tip" my hat to both Andy and Elliot for giving me the opportunity to change teams, even if only for this essay.

When I was teaching at Warren College at UCSD back in the mid and late 1980s it was not uncommon for several colleagues to discuss the current scientific ideas of the day. Paul and Patricia Churchland, the famous philosophers of mind, had just arrived on campus, and the fledging field of cognitive science was finding its bearings. One afternoon, the Director of the Writing Program at Warren College, James Crosswhite (now Director of the Composition Program at the University of Oregon) and I got into a friendly debate about consciousness and materialism. He more or less took a materialist position and, I, because of a deep interest in Indian spirituality, assumed an idealist posture that mirrored some of Ramana Maharshi's Advaita Vedanta arguments.

Given my youth and exuberance for a Consciousness first ideology I felt fairly certain I had won the argument that day. However, given Dr. Crosswhite's acute intelligence, I am also fairly certain that my recollections are probably wrong.

36

But I do recall one line of reasoning that stood out and which appeared to all who were present (a number of colleagues had gathered around Crosswhite and me as we engaged over this eternal philosophical question) to be at least of some merit, if only as a sharp analogy.

Here is how the thought experiment goes: Imagine that the only state of consciousness that existed was, in fact, our dream world. Further imagine that in such a state an unusual person (we will call him Ramana) confronts you by claiming to have access to a hitherto unknown level of awareness, which he calls the "waking" world. Ramana further argues that all experiences within the dream state are subsumed (indeed produced) by a waking brain which is inaccessible to dreamers. And therefore the attempt by the majority of dream materialists to reduce waking phenomena down to their dream stuff is completely wrong and misleading, since the truth of the situation is completely the opposite. The dream is happening because of the waking state brain (in another realm) not the apparent dream brain that looks to be generating awareness from itself and from its extended environment. Ramana's ultimate point is a very simple one. While it may seem to overwhelmingly clear that the dream brain causes the dream world, the fact is that a transcendent state of being is its real cause and origination.

What to make of such a claim? If one were grounded in "dream stuff" methodologies, one might ask for some convincing evidence of such a world (let's call this thinker Churchland). To which Ramana might reply that it was impossible to actually transport such waking stuff into the dream world, since the very moment one attempts to do so it instantly transforms into dream material. Churchland, ever the skeptic, might then rejoinder that Ramana's bold claims lack proof and as such warrant no further attention. Or, she might argue that Ramana's waking excursions were just modifications of his own dream brain and that what he thought was higher and transcendent was neither, since his numinous experiences were the result of neural dream discharges within his own dream skull.

At this juncture, Ramana may argue that to see the proof of his claim one must be willing to do a most radical experiment. One must literally "die" to the dream state in order to properly access the waking state. When that happens, then Churchland will actually have the extraordinary evidence she demands. Of course, Churchland might balk at this suggestion since dream death seems a bit extreme to prove a point.

Churchland may persist and query Ramana again and say, "Why can't you produce something in our present state of awareness which would give us confidence that your claims are true?" In addition, is it not possible that you are wrong, Ramana, since your recollection of the waking state must be recalled in and through your present dream brain? How do you possibly know that that the dream brain couldn't produce what you wrongly believe is higher?

Ramana may then point out that the certainty of any experiment rests upon an uninspected axiom: that the present form of awareness is somehow the best and final arbitrator of all other states of consciousness. Why this is so isn't an ontological fact and if other super-luminal forms of awareness do exist then exploring them may help us contextualize our present dream state.

At this stage then Ramana could encourage Churchland to take up the experiment by practicing a method that he himself used. Using one's own self-awareness ask what is the source of such luminosity. According to Ramana, that very inquiry will lead to a deep questioning of what one takes to be "real" and "permanent" and will eventually prompt one to emerge into the waking state, which itself is the larger context behind the dream world. In such an awakening, the erstwhile skeptic will immediately discover that dream stuff was not the real cause behind dreaming. Rather, it was a physical brain in a completely different state of awareness.

I didn't exactly argue like this with Dr. Crosswhite. Rather, I merely took Ramana's consciousness as first principle and argued accordingly, borrowing and creating thought analogies when appropriate.

38

The gist of my argument was that saying everything was caused by the physical brain may appear to be perfectly sensible in this present waking state awareness, but may in truth be completely wrong if indeed there were higher states of awareness in which this and other states were subsumed.

More precisely, even if a purely materialist position absolutely convinced us that matter and only matter gave rise to our self-awareness, we could still be quite mistaken. The dream analogy or Fritz Staal's *Three Ants in a Room* analogy or Plato's *Allegory of the Cave* or Edwin Abbot's *Flatland* all address, to varying degrees, the idea of a multi-dimensionality to consciousness. The waking state does indeed seem certain until we fall asleep. Likewise, a lucid dream appears real until we wake up. If that is the case with two states, is it really that unreasonable to think of a third or a fourth state that would show the relativity of our present state?

I think I was convinced of Ramana's reasoning and the efficacy of the interior search from an early age (I first read Yogananda's *Autobiography of a Yogi* when I was eleven years ago) and because of that was well acquainted with how certain schools of Indian philosophy might counteract skeptical inquiries. I started practicing yoga and meditation when I was barely 12 years old so I had some acquaintance with the varying techniques and methods that certain Indian school advocated. I was also quite familiar with monastic spirituality in Christianity and Roman Catholicism, especially the more mystical writings of Nicholas of Cusa, St. Teresa of Avila, St. John of the Cross, Richard Rolle, Walter Hilton, and the pseudonymous author of the Russian classic, *The Way of a Pilgrim.*

But arguing with conviction (and I certainly was convinced by my own personal epiphanies) is also a pathway fraught with all sorts of dangers, not the least of which is conflating one's mystical encounters with ultimate truths.

This hubris in me became more apparent when I had a prolonged debate via proxy with Patricia Churchland, Professor of Philosophy at UCSD and one of the creative architects behind neurophilosophy (her book of the same title has been seminal in the field). It turned out that one of my students in my Warren College class at UCSD was also taking a course with Patricia Churchland that focused on the neurobiological basis of awareness. Tom Wegener and I had by chance met earlier while surfing a point break in Carlsbad and we were both pleasantly surprised to meet again in an academic setting. Tom was a philosophy major at the time and was keenly interested in Churchland's eliminative materialism, whereby old folk-psychological tales are replaced by more accurate and robust scientific explanations. Invariably Tom would walk into my classroom and tell me about the latest insight he learned from Dr. Churchland's brilliant mind. In my arrogance, I would immediately counter her materialism with my more idealistic Ramana musings, trying to convince Tom that even though neurobiology makes eminently good sense, particularly when applied to philosophy, it doesn't mean that consciousness itself can be absolutely reduced to a set of neurons firing. [*Sidenote*: After graduating from UCSD and then getting his law degree from USD, Tom eventually moved to Australia where is now builds surfboards out of *alaia* wood like the ancient Hawaiians.]

Of course, this wasn't precisely what Patricia Churchland was saying, but in my own form of reductionism it was the easiest way for me to underline what was at stake. Tom would then go from my class to hers (or was it vice versa?) and then attempt to get Professor Churchland's opinion of this non-materialist explanation of the mind. Naturally, she didn't buy it for a second, and would rebut my varying points with her usual alacrity.

However, her gifted retorts didn't fundamentally change my position, despite the fact that I very deeply admired Patricia and her husband-colleague, Paul, who was also a Professor in the department.

Why was I so hardheaded? Well, I am sure there are several reasons for my stubbornness (ranging from epigenetic recourses to hidden brain lesions), but I think the overriding reason is because I, myself, knew from my own experiences that when one is in a higher state of awareness it makes the waking state look like a dismal and shadowy dream. It seemed as if neuroscience was merely the logical extension of Flatland thinking, of taking the dream brain too seriously, and snuffing out the possibility that deep meditation or self-inquiry could actually lead beyond the rational mind into realms thought impossible by physicalist thinking.

Of course, a whole host of thinkers have postulated this as well, ranging from early Gnostics to modern day mathematicians and quantum physicists. This is perhaps one of the reasons why Ken Wilber is so popular among a wide of range of readers. He is in a long line of philosophers who have advocated that there is something beyond our five senses, even if what is "beyond" may be deeply embedded within such material constructs.

But one of the great difficulties confronting such perennialist thinking (and this includes even present-day Wilber, even if he intimates otherwise) is that arguing by analogy or by experiential inquiry doesn't easily translate as "evidence" or "proof" in the nuts and bolts world of everyday life.

Yes, there seems to be little question that amazing states of consciousness exist and that almost all humans can access such under certain conditions. Even the most skeptical of scientists will concede the plasticity of awareness. But such a phenomenological acceptance doesn't mean that we have agreed upon what these experiences actually mean or portend. The impasse between materialists and mystics isn't over whether UOE's (unified oneness experiences) or NDE's or OBE's or any other illuminated "E" experience exist, but how to best interpret and explain them. In other words, how do we decide or know that consciousness isn't reducible to the known laws of physics and neuroscience? We have already reduced water to its chemical make-up of H20 and nobody seems too concerned that we have more or less eliminated Neptune as a guiding explanatory principle. One could argue that science is one long (and quite successful) history of what occurs when humans discover a physical explanation for what was otherwise regarded as the province of god or metaphysics.

41

If this has been the case for explaining almost the entire known universe, from electromagnetism to gravity, why shouldn't science also be successful in explaining human awareness? And aren't mystics and spiritualists and religionists too prone to explain their numinous encounters with outdated modes of thinking? As I argued in the *Politics of Mysticism*:

But, the argument goes, the devoted mystic will say that his or her experiences are authentic (because of the utter certainty of the encounter) and the experiences of others, especially if they belong to a rival group which splintered off after a succession dispute, are misguided, secondary, or illusory. So what we actually have in effect here in in terms of truth claims is not essentially different than that of a fundamentalist. The mystic is right by virtue of his/her inner attainment and everybody else is wrong (no matter how politely we may gloss over it: karma or chance?) because he/she happened to get the right guru and the right path (and by right we mean "highest").

But notice how the mystic is not calling into question or doubt his/her own truth claims. For example, one rarely finds a completely agnostic posture among disciples about the relative status of his/her guru. Why not? Because just like the fundamentalist he or she is not trained to severely doubt interior revelations of truth, primarily because they appear so real when they occur. It is one thing to state that my inner experiences have convinced me that I am on the right track; quite another to then make judgments on the veracity of other meditators' experiences.

To strike a sociological note here, it becomes fairly apparent that culture plays a significant role in the ultimate interpretations of inner experiences. What at first glance appears to be a simple, sweet path to enlightenment, turns out to be on closer inspection a political contest over religious claims--claims, I should add, that have been transformed by the cultural landscape of when and where they take place. We may wish that mysticism was devoid of culture, or personal bias, or religious prejudice, but it is almost wholly entrenched in it.

Why? Because we never apprehend inner lights and sounds and beings divorced of their interpretative network. In other words, our socially conditioned minds are always flavoring, always transforming, and always contextualizing whatever we perceive, whether those sights are inner or outer. And it is exactly when my experiences are personal and internal that I am most subject to error. Why? Because we have yet to discern a normative corrective for mystical encounters. Sure we have templates to gauge inner experiences, their relative efficacy and so on, but since most individuals have no mastery of leaving their bodies we are subject to tremendous imprecision and tremendous speculation. Yet do we admit to this impasse? Do we acknowledge our immaturity in the spiritual arena?

There is something fundamentally skewed when religious converts (of any persuasion and of any methodological bent) begin to believe that they have cornered the market on truth. As one wise saying puts it, "If there really is a God, He/She may find atheism to be less of an insult than religion." The point is obvious: what we know the least about is the very thing we make absolute statements on. Strange, but true. Take Jesus Christ, as a prime, if controversial, example. What do we really know about him? Not very much. Depending on your perspective and the sources that you cite, Jesus emerges as the only begotten Son of God, a Jewish mystic with Gnostic leanings, or a clever, but ultimately misguided magician. The only thing that is absolutely certain about Jesus, at least historically speaking, is that we know less about him than we think. Indeed, the real truth about Jesus' existence is forever buried in the recesses of time.

And yet we have some one billion plus people on this planet right now who more or less believe that if you don't accept the truth claims of Jesus Christ you will end up in eternal hell. All of this and we still don't know what he even looked like and what he did for some fifteen years in his teens and early twenties? Couple this with the contradictory and entirely insufficient biographical details contained in the gospels which are the major sources for Jesus' life and you wonder how a Christian can be so assured in their faith. Put bluntly, you wouldn't allow your son or daughter to marry a prospective suitor if the only information you had on them was equal to what we know about Jesus. But there are millions of us who seriously think that we have to make a lifetime, nay eternal, commitment to a person we have never met and know less about than our next-door neighbor.

When it comes to religion and its claims, whether they are based on revealed texts or interior visions, the one common denominator is that we somehow have to check our brains in at the door before entering into the tabernacle of ultimate truth. Yet it is exactly that brain, that three pounds of wonder tissue or glorious meat, as Patricia Churchland so succinctly puts it that has allowed us to

43

ponder life's ultimate questions. It is that very brain which has led us to pray, to read, to meditate. It is also that very brain which can misinterpret exterior stimuli as well as internal neural firing. My hunch is that before we make any ultimate claims for truth, we understand that we are constantly subject to error.

So the mystic may potentially be better off than the mere believer, who only reads but never actually engages in technical spiritual practices, because he or she gets firsthand experiences of alternate realms of consciousness not merely menu descriptions of them. But this does not mean that the mystic has experienced the "truth" in all its purity and that the mystic somehow "knows" the efficacy of other spiritual teachers or paths. No, what the mystic does in fact know is rather quite simple: a different state of consciousness which he or she interprets according to his/her cultural or religious background. On that score, I do think that mystics are on the right track; it is better to experiment than simply speculate. Yet, the results of those experiments are subject to numerous interpretations, some of which are better than others. Since we are still at such a preliminary level in our investigation of states of consciousness beyond the waking-rational level, it seems to me to be a much wiser course for us to adopt a stance of honest humility and openness than succumb prematurely to absolute statements or theorizing which in the end causes much more harm than good.

But isn't the materialist agenda too myopic for its own good? John Searle, Professor of Philosophy at U.C. Berkeley, persuasively argues that third person descriptions of first person narratives cannot adequately do justice to the subjective nature of such experiences.

And even if analogies cannot constitute evidence, they can at the very least prompt unexpected voyages which can on their return trips provide the tangible evidence that was missing at the outset--witness Charles Darwin and his five year journey on the *H.M.S. Beagle* or Captain Cook's encounters in Tahiti and beyond.

What kinds of evidence can a mystic proffer that would convince neuroscientists that their very paradigm may need to be transcended?

Perhaps the evidence we seek is by its nature transcendent and not amenable to empirical test claims? If so, then we are truly not in a Newtonian or Einsteinian world anymore and we should straight up admit it. That is, if mysticism is indeed a "transpersonal" science then it may just have to go it alone and forget convincing us flatlanders otherwise. I say this because if consciousness is indeed multi-dimensional in an ontological sense then it won't be possible to reduce one state to another without concealing its most important features.

If this is indeed the case, as some mystics have argued, then we may be confronting the limits of what has been tantalizingly termed the *Chandian Effect*. It was so named because Faqir Chand was the first Sant Mat guru to speak at length about the "unknowing" aspects of visionary manifestations. In this context, the *Chandian Effect* designates two major factors in transpersonal encounters: 1) the overwhelming experience of certainty (*ganz andere* and *mysterium tremendum*) which accompanies religious ecstasies; and 2) the subjective projection of sacred forms and scenes by a meditator/devotee without the conscious knowledge of the object/person who is beheld as the center of the experience. The *Chandian Effect* in the realm of mystical experiences is weakly analogous to Heisenberg's principle of uncertainty in subatomic physics. The more "certain" or "real" the mystical encounter seems, the less likely one is to believe that such is the product of subjective projection or transference. This invariably causes deep epistemological consternation, since what makes us certain that something is indeed real is the result of our own deeply felt subjectivity (even if dressed up in objectivist language).

This explains, albeit only partially, the great transformation that occurred to Ramana Maharshi of South India. Paul Brunton in his book *A Search in Secret India* retells it: "He [Ramana] was sitting alone one day in his room when a sudden and inexplicable fear of death took hold of him. He became acutely aware he was going to die, although outwardly he was in good health. He stretched his body prone upon the floor, fixed his limbs in the rigidity of a corpse, closed his eyes and mouth... 'Well, then,' said I to myself, 'this body is dead. It will be carried stiff to the burning ground and then reduced to ashes. But with the death of the

45

body, am I dead? Is the body I? This body is now silent and stiff. But I continue to feel the full force of my self apart from its condition.' These are the words with which the Maharishee [Maharshi] described the weird experience through which he passed... He seemed to fall into a profound conscious trance where in he became merged into the very source of selfhood, the very essence of being. He understood quite clearly that the body was a thing apart and that the I remained untouched by death. The true self was very real, but it was so deep down in man's nature that hitherto he had ignored it."

As Ramana himself gracefully said: "There is only one Consciousness and this, when it identifies itself with the body, projects itself through the eyes and sees the surrounding objects. The individual is limited to the waking state; he expects to see something different and expects the authority of his senses. He will not admit that he who sees the objects seen and the act of seeing are all manifestations of the same Consciousness-the 'I-I' [Real Self]. Meditation helps to overcome the illusion that the Self is something to see. Actually there is nothing to see. How do you recognize yourself now? Do you have to hold a mirror up in front of yourself to recognize yourself? The awareness is itself the 'I.' Realize it and that is he truth."

This mystical resonance is quite understandable since the experience brings forth its own definite and convincing certification, just as the waking state does after a good night's sleep.

But herein arises a pertinent observation. Our conviction that something is real or certain doesn't mean that we cannot be mistaken. We most definitely can be.

A good example of this comes from Faqir Chand's own life story, published near the end of his life in his autobiography, *The Unknowing Sage*, and also visually explained in the short film, *Inner Visions and Running Trains*. Writes Faqir:

In 1919 1 was posted in Iraq. The aboriginal inhabitants (known as Baddus) revolted, which led to a fierce battle, I was inspector in the department of telegraphy in the railways with my headquarters at Divinia. The rebels made a heavy attack on the Hamidia railway station, killing the entire staff and setting the building on fire.

Military forces from my post were rushed to Hamidia. I was also ordered to take charge of the Hamidia rail- way stations as Station Master. Our soldiers laid down wires in trenches and occupied their positions. Fierce fighting continued and there was a heavy loss of life on both sides. At Hamidia we were left with a group of thirty-five soldiers and one Subedar Major. The rest of the army was sent to Divinia to retaliate anyattack there. With the fall of the night the rebels attacked us. Our sol- diers, though less in number, fought back. One of our men was wounded while casualties on the opposition were very heavy. As the firing ceased for some time, the Subedar Major came to me and asked that I convey to our headquarters at Divinia that we were short of ammunition. And, if we had to face another such attack, our supplies would not last for more than an hour. If the ammunition supply failed to reach us before dawn, none of us would be alive. I wired the message to the headquarters accord- ingly. The situation was tense and everybody was feeling as if the end had come. I too was shaken with the fear of death. In this very moment of fear the Holy Form of Hazur Data Dayal Ji appeared before me and said, "Faqir, worry not, the enemy has not come to attack but to take away their dead. Let them do that. Don't waste your ammunition." I sent for the Subedar Major and told him about the appearance of my Guru and his directions concerning the enemy. The Subedar Major followed the directions of my Guru. The rebel Jawans came and carried away their dead without attacking our positions. By six o'clock in the morning our airplanes came and they dropped the necessary supplies. Our fears vanished. We gained courage. We were safe.

After about three months the fighting came to an end and the Jawans retired to their barracks. I returned to Baghdad, where there were many satsangis. When they learned of my arrival, they all came together to see me. They made me sit on a raised platform, offered flowers, and worshipped me. It was all very unexpected and a surprising scene for me. I asked them, "Our Guru Maharaj is at Lahore. I am not your Guru. Why do you worship me?" They replied in unison, "On the battle field we were in danger. Death lurked over our heads. You appeared before us in those moments of danger and gave us direction for our safety. We followed your instructions and thus were saved." I was wonder struck by this surprising explanation of theirs. I had no knowledge of their trouble. I, myself, being in danger during those days of combat, had not even remembered them.

This incident caused me to question within myself, "Who appeared to them? Was it Faqir Chand?"

47

What Faqir Chand eventually realized was that what was appearing within both himself and his colleagues was a projection of his own mind.

Faqir Chand explains:

Now, after having such experiences with me, I question myself, "Faqir Chand, say, what mode of preaching do you wish to change? Which teachings should I change?" The change that I can make in the present mode of preaching I explained in the discourses that I delivered during my tour. The change is, "O man, your real helper is your own Self and your own Faith, but you are badly mistaken and believe that somebody from without comes to help you. No Hazrat Mohammed, no Lord Rama, Lord Krishna or any God or goddess or Guru comes from without. This entire game is that of your impressions and suggestions which are ingrained upon your mind through your eyes and ears and of your Faith and Belief." This is the change that I am ordained to bring about.

Faqir Chand came to realize that his numinous experiences were not indications of a transcendent state of consciousness (though during the experiences it seemed to be exactly that), but was rather projections of his own mind.

When he began to doubt the reality of these visions he broke through into another state of awareness which itself seemed more luminous than the stage preceding it. But even this new ascending form of consciousness he believed was also a projection.

But the more certain the experience the less likely it was to reveal its real origins. In other words, the very illumination blinded one from discovering its underlying cause. It was only when one "doubted" what one saw and heard and experienced that one was able to wiggle free from its overwhelming certainty.

Faqir summarized this best near the end of his life, after nearly 70 years of continuous deep meditation: "So what I have understood about Nam is that it is the true knowledge of the feelings, visions, and images that are seen within. This knowledge is that all the creations of the waking, dreaming and deep sleep modes of consciousness are nothing but samskaras (impressions which are in truth unreal) that are produced by the mind. What to speak about others, even I am not aware of my own Self (in dreams). Who knows what may happen to me at the time of death? I may enter the state of unconsciousness, enter the state of dreams and see railway trains. . . How can I make a claim about my attainment of the Ultimate? The truth is that I know nothing."

If each state of consciousness tends to blind us about its causation then determining what is a higher or lower state seems predicated upon whatever state we are presently in. Is there any way out of this epistemological *cul de sac*?

Ramachandran in a series of groundbreaking papers demonstrates how easily the brain can deceive our perceptions of physical realities. In *Brain: The Journal of Neurology* (Volume 121, number 9) Ramachandran writes:

"Almost everyone who has a limb amputated will experience a phantom limb —the vivid impression that the limb is not only still present, but in some cases, painful. There is now a wealth of empirical evidence demonstrating changes in cortical topography in primates following deafferentation or amputation, and this review will attempt to relate these in a systematic way to the clinical phenomenology of phantom limbs. With the advent of non-invasive imaging techniques such as MEG (magnetoencephalogram) and functional MRI, topographical reorganization can also be demonstrated in humans, so that it is now possible to track perceptual changes and changes in cortical topography in individual patients. We suggest, therefore, that these patients provide a valuable opportunity not only for exploring neural plasticity in the adult human brain but also for understanding the relationship between the activity of sensory neurons and conscious experience. We conclude with a theory of phantom limbs, some striking demonstrations of phantoms induced in normal subjects, and some remarks about the relevance of these phenomena to the question of how the brain constructs a 'body image.'"

If we can anatomically be mistaken when it comes to ordinary reality, which we can easily double check, then it seems we are confronted with much larger issues of literal confusion when it comes to alternative states of consciousness.

Of course, Michael Shermer, founder of *Skeptic Magazine,* critiques the natural/supernatural divide from a different angle:
"I don't think a union between science and religion is possible for a logical reason, but by this same logic I conclude that science cannot contradict religion. Here's why: A is A. Reality is real. To attempt to use nature to prove the supernatural is a violation of A is A. It is an attempt to make reality unreal. A cannot also be non-A. Nature cannot also be non-Nature. Naturalism cannot also be supernaturalism. In a natural worldview, there is no non-natural or supernatural. There is only the natural and mysteries left to explain through natural means. Believers can have both religion and science as long as there is no attempt to make A non-A, to make reality unreal, to turn naturalism into supernaturalism. The only way to do this for theists is to posit that God is outside of time and space; that is, God is beyond nature—super nature, or supernatural—and therefore cannot be explained by natural causes. This places the God question outside the realm of science. Thus, there can be no conflict between science and religion, unless one attempts to bring God into our time and space, which is a violation of the principle of A is A."

If we substitute mysticism or transcendent consciousness for religion, Shermer's argument is that there is an indissoluble gap between science and spirituality. And any attempt to bridge the two is a violation of the simple principle that A is A. While I am not quite convinced by Shermer's verbal sleight of hand, I do think he is on to something that deserves closer inspection.

For instance, if consciousness is indeed brain based then neurology should be able to help us better understand how it arises from material structures. Mysticism or the pursuit of higher states of consciousness, therefore, will also be part and parcel of such neurological studies, and will not be beyond its jurisdiction. However, if consciousness (or some part of our awareness) is not physically produced then science will not be able to comprehensively explain it as such. This implies that science will eventually confront a border it cannot cross.

Now scratch out everything I have just written and let's talk about a new application on the iPhone called the "sound grenade."

Okay, before you think that I have just forced you into one very strange non sequitur (or, what I like to call a hypertext parenthetical), it may well be that the real difficulty in studying consciousness can be easily demonstrated without any words whatsoever.

My son Shaun downloaded this application that plays (I am told) one very annoying high pitched sound which can be quite irritating. Some commentators have even mentioned getting sick to their stomachs after hearing it.

Well, Shaun likes to randomly explode his sound grenade while eating dinner or watching television to both surprise and annoy his young brother Kelly, and his mother, Andrea.

However, it never works on me. Why? I cannot hear it. In fact, I hear absolutely nothing when he turns it on. At first I thought my family was in on some secret joke, since I didn't believe that the application was really emitting a sound. Later when I was convinced that they were not lying I was a bit wonder struck. Why do I hear absolutely nothing?

So, one day when I was playing Frisbee golf I secretly turned on the application to see if I could annoy my brother, Joseph, and throw him off as he went for a birdie. Nothing happened. I tried it again and again nothing happened. I finally told him about the high-pitched sound that was supposedly generated by the application and he couldn't hear a thing. In fact, he too didn't believe that the device really emitted a sound and he proceeded to get a bit irritated with me as he thought I was playing some stupid joke on him. Even to this day, I still don't think he believes that the sound grenade really does work.

The other day I thought I would test the application on my Religion and Science students at CSULB. Everyone heard it except me and one other student, who was also flummoxed to be the only one of his colleagues to not hear it.

Why do I bring up this apparently silly example? Because I think it strikes at the very heart of the inherent difficulties in consciousness studies. For instance, how can one properly

51

study the physics of the sound grenade application on the iPhone if the one studying it cannot hear what everyone else is hearing? Correlatively, how can one study higher states of consciousness if one has never experienced them?

I understand that it might be theoretically possible to do so, but doesn't trying to apprehend a given phenomenon by way of a surrogate versus one's own immersion fundamentally lose something? *Do we really think that someone who is not conscious can truly understand one who is?*

Now, ironically, the sound grenade analogy doesn't mean by way of extension that higher forms of awareness are metaphysical (the iPhone application uses real sound waves, not astrally generated ones), but only that a spectrum of variances may exist that shouldn't ad hoc be collapsed to each other.

If we prematurely do so, we end up with what Daniel Dennett rightly called "cheap" reductionism. It may look valuable but on closer inspection it offers us nothing useful in exchange.

To say my consciousness is merely the result of a bundle of neurons is neither enlightening nor useful. What we really want to know is how such a set of tiny physical on and off points could produce self-reflective awareness. This is a technical problem, not a philosophical one.

And it is for this reason that scientists such as Gerald Edelman have tried to see if it is possible to construct artifacts that are self-aware. As Edelman explained in an interview with *Elmundo Digital*:

Question: Do you believe that it will be possible to create robots that replicate the working of the brain in the future?

Answer: This is just what we are doing in my lab. We are trying to create a conscious machine. In fact, we have already built devices whose performance is based on the structure of the brain. They look like robots, but I wouldn't call them that way, because they don't have an automatic programmed behavior, they have an artificial brain whose design in based on what we know about the human brain structure. These devices, even not being living entities, are able to perform some cognitive operations that imply the usage of memory.

Question: For instance?

Answer: They can learn by heart different paths to an object, and apply learning to get to the object by the shortest path. In fact, our devices have participated in robot soccer tournaments, and they have won all the matches because they are able to learn and adopt strategies. In sum, today we can say that we have managed to build devices that are able to do certain things by themselves; this is something that 10 years ago I myself would have said to be science fiction. Therefore, nowadays I would dare to say that, once we understand more about the structure of the brain, we will be able to build conscious machines in the future.

Yet, on the other hand, the conviction of my own experiences (or lack thereof) is not an indication of its causation. Rather, we have an almost innate predisposition to confuse our own transferences and projections for objective realities or truth--neglecting in the process just how such numinous illuminations arise in the first place.

The problem with mysticism, therefore, isn't the lack of subjective experiences of extraordinary realms but the inability while in such exalted states to recognize the humble and ordinary bases for their generation.

This confusion of a neural system about its own interpretation of reality can be quite startling. As Gerald Edelman recounts, "There's a neurologist at the University of Milan in Italy named Edoardo Bisiach who's an expert on a neuropsychological disorder known as anosognosia . A patient with anosognosia often has had a stroke in the right side, in the parietal cortex. That patient will have what we call hemineglect. He or she cannot pay attention to the left side of the world and is unaware of that fact. Shaves on one side. Draws half a house, not the whole house, et cetera. Bisiach had one patient who had this. The patient was intelligent. He was verbal. And Bisiach said to him, "Here are two cubes. I'll

53

put one in your left hand and one in my left hand. You do what I do." And he went through a motion. And the patient said, "OK, doc. I did it." Bisiach said, "No, you didn't."

Of course, the problem for the scientist may be exactly the opposite—the inability while working in an ordinary state of awareness to recognize the superluminal bases for its very existence.

Perhaps the study of consciousness can benefit by listening more carefully to that ancient quip about spirit and matter. "The more I study the mystical, the more physical it becomes and the more I study the physical, the more mystical it turns out."

"But, strange to say, although she had so made up her mind not to be influenced by her father's views, not to let him into her inmost sanctuary, she felt that the heavenly image of Madame Stahl, which she had carried for a whole month in her heart, had vanished, never to return, just as the fantastic figure made up of some clothes thrown down at random vanishes when one sees that it is only some garment lying there. All that was left was a woman with short legs, who lay down because she had a bad figure, and worried patient Varenka for not arranging her rug to her liking. And by no effort of the imagination could Kitty bring back the former Madame Stahl."

—Leo Tolstoy, Anna Karenina

4 | Tangled Phone Lines: Dawkins vs. Wilber

Contrary to what Elliot Benjamin in "The Boundaries of Science" took away from my recent article, *Is My iPhone Conscious?* I clearly believe that mysticism can be studied scientifically. Indeed, it already has been for decades.

I don't necessarily think that mystics and skeptics should depart company and go on their respective ways, but I do think that if they seriously communicate with each other something is bound to give. And what each side may have to give up is more than they might be willing to concede.

For example, take a mystic in the Sant Mat tradition, such as the late Kirpal Singh (who Ken Wilber once cited as the "unsurpassed master of the subtle realm"), who practiced an ancient yogic technique known as shabd yoga wherein one listens to a divine inner sound which reportedly transports the soul into higher regions of bliss beyond the body and mind. The theological framework undergirding much of Kirpal Singh's practice (which dates back centuries through such practitioners as Tulsi Sahib, Dadu, Kabir, and Guru Nanak) states emphatically that the shabd (or inner sound) is not physical or brain produced but is rather a divine melody that transcends all of physics and this universe.

But what happens if in studying this phenomenon from a neurological point of view the Sant Mat mystic discovers that such "sacred" music is tinnitus, which may be caused by earwax buildup, or nasal allergies, or lower levels of serotonin, or any host of physically related ailments. Does this sort of information change the numinous encounter or its interpretation?

Are religious movements, predicated on shabd yoga practice, such as Radhasoami, the Divine Light Mission, Eckankar, Quan Yin, MSIA, etc., going to accept such a reductionistic explanation of what erstwhile was regarded as a direct pathway to God-Realization?

I think not.

Why? Because the very basis of the practice is centered on the notion that it is a spiritual endeavor and not merely a cranial one.

In other words, if a mystic is serious about studying the subject scientifically it means that he or she may have to radically revise their understandings and prior theological dogmas about what is actually happening when they undergo a transformation of consciousness.

This is not to suggest that mystics haven't done so in the past. They have, but this tends to be the exception to the rule not the norm.

The best example that I know of personally comes from Faqir Chand who originally believed that the inner visions he had of Krishna and his human guru, Maharishi Shiv Brat Lal, were objectively real. Only later when Faqir became a guru in his own right and started hearing reports from his disciples that he appeared in visions to them (without any conscious knowledge on his part) did he doubt their exterior reality. This led Faqir Chand to radically change the mode of his teaching.

After meeting personally with Baba Faqir Chand, it became exceedingly apparent to myself and Professor Mark Juergensmeyer (who visited Manavta Mandir in late August of 1978. See Juergensmeyer's book *Radhasoami Reality* (Princeton University Press, 1991) that the old sage was something of an anomaly amongst Indian gurus. For, although Faqir Chand had a rather large and devoted following (numbering in the thousands), he absolutely disclaimed himself of any miracles attributed to his spiritual work, saying quite frankly that they were products of either the devotee's previous karma or intense faith. Indeed, it was this very insight that led Faqir to his own Enlightenment.

When Faqir Chand began to initiate disciples into *surat shabd yoga*, at the request of his master Shiv Brat Lal, a most curious thing happened. His devotees began reporting that Faqir's radiant form appeared inside their meditations. Others related miracles that were caused by Faqir's *prashad* (blessed food), letters, or advice. However, all during this time Faqir claims that he had absolutely no knowledge or awareness of his form appearing to distant provinces or performing miracles to the sick and dying. As Faqir himself wrote: "People say that my Form manifests to them and helps them in solving their worldly as well as mental problems, but I do not go anywhere, nor do I know about such miraculous instances." It was at this point when Faqir asked himself: "What about the visions that appear to me? Are they a creation of my own mind, and does my guru also not know about his appearances to me?"

Only then, according to Faqir, did he realize the truth: "All manifestations, visions, and forms that are seen within are mental (illusory) creations."

After his realization, Faqir began preaching his belief that all saints, from Buddha, Christ, to even his own master Shiv Brat Lal are ignorant about the miracles or inner experiences attributed to them. In a paper given to the American Academy of Religion in March 1981, I used the term "The Unknowing Hierophany" to describe what Faqir Chand believes; that is, a "Divine" vehicle within the temporal world that is unaware of its spiritual manifestations. A revised form of this original paper was published under the title "The Hierarchical Structure of Religious Visions," in *The Journal Of Transpersonal Psychology* (Volume 15, Number 1).

What is perhaps most striking about the Faqir Chand example is how he was viewed by other gurus in the Radhasoami tradition. Instead of incorporating his findings into a new and updated understanding of what occurs during meditation, most of the spiritual leaders either ignored him or dismissed him as senile (which, of course, is ironic given that Faqir "doubted" his subsequent visions instead of believing them).

The other Radhasoami mystics didn't want to "give" on this issue of visions, since apparently conceding that all such manifestations were merely illusions or projections of the disciple's own mind (and that the initiating guru had nothing to do with the mysterious phenomena) would undermine their very tenure.

If there is going to be a rigorous science of mysticism, one of the first casualties will be the superstructure (as Frits Staal, my old professor at U.C. Berkeley, termed it) around it. In other words, the more science is allowed to understand and explain mysticism the less mysterious and metaphysical it will appear.

Indeed, the history of science is in many ways the elimination of formerly believed gods. We replace a phlogiston theory with an oxygen theory, realizing in the process that phlogiston never existed in the first place. We didn't explain phlogiston; we eliminated it because it was wrong.

Likewise, the scientific study of mysticism may replace some deeply cherished ideas. Some which may be so cherished that their very elimination will be too heavy a price for some.

Francis Crick has already suggested that the soul will never be explained for the simple reason that it never existed.

As Crick so astutely puts it: "The Astonishing Hypothesis is that 'You," your joys and your sorrows, your memories and your ambitions, your sense of personal identity and free will, are in fact no more than the behavior of a vast assembly of nerve cells and their associated molecules.' As Lewis Carroll's Alice might have phrased it: 'you're nothing but a pack of neurons.' This hypothesis is so alien to the ideas of most people alive today that it can be truly called astonishing."

Now on the surface of it, Crick's argument seems so obvious as not to be very astonishing at all, especially when we realize that every great discovery in science has been grounded, so to say, in some simpler material structure. Cells turned out be cast from molecules; molecules from atoms; atoms from electrons, protons, and neutrons. Even questions as profound as *What is Life?* (also the title of a highly influential book by the famed physicist Erwin Schrodinger) which plagued biologists during the early and mid part of this century turned out to have a physical, if minute, answer: the D.N.A. molecule. Yet, many thought the answer would never be found because life was something vitalistic, something non-material, and something science could not identify. As it turns out, though, every endeavor to locate the secrets of the universe hinge on focusing first and foremost on the empirical realm. As Crick sees it, why shouldn't consciousness have a physical basis? Hearing does. Seeing does. Why not being as well?

And even closer to our own secular bone, we may well discover that our "self" doesn't even exist. As a summary of Thomas Metzinger's recent book, *The Ego Tunnel*, puts it:

"For Metzinger, the conscious self is really nothing more than the content of a "transparent self-model" – an image of ourselves in the brain, an image that we are unable to recognize as a model. It is from this image that an Ego emerges.... The Ego Tunnel surveys a range of findings about the deep structure of that conscious self. For instance, Metzinger examines how people born without arms or legs can experience a realistic sensation that they do in fact have limbs. He explores both the workings of body and self in the dream state, and our control of those states to experience the effects of lucid dreaming. Further, he recounts experiments that demonstrate how out-of-body experiences and sensations of action and agency can be induced by directly stimulating the brain. From studies of long-term meditators, too, he gives us new insights about the ways in which the unity of consciousness is constituted in the brain. Working from these examples and more, Metzinger argues that "the" self of subjective experience is actually created by our brain mechanisms. Now, as new ways of manipulating the conscious mind-brain appear on the scene, it will soon become possible to alter our subjective reality in an unprecedented manner. The cultural consequences of this, Metzinger claims, may be immense: we will need a new approach to ethics, and we will be forced to think about ourselves in a fundamentally new way. If actions are really the product of brain states, as opposed to intentions, then what, exactly, is the nature of free will? And if one day we can create artificial systems that generate conscious selves—Ego Machines—should we actually do it? The groundbreaking findings outlined in The Ego Tunnel have implications for how we think about drug use and the brain; the ethics of neurotechnology; the extent of personal responsibility; and a host of other contentious issues. Metzinger ultimately argues that we must be willing to engage with the serious moral and cultural questions that will arise in the wake of this new image of humanity. At a time when the science of cognition is becoming as controversial as the theory of evolution, The Ego Tunnel offers an accessible introduction to the field of consciousness studies, a new theoretical vision and a compelling perspective on the mystery of the mind."

To put this into sharper relief, as one member of Radhasoami once complained to me, "Why should I meditate for 3 hours a day if what I am experiencing isn't God but sophisticated neural fireworks."?

In other words, isn't at least a good part of the mystic quest predicated upon a false idea to start with? We aren't looking for just stuff, as Patricia Churchland once put it; we are looking for some divine meaning.

I think there is a reason Ken Wilber and Richard Dawkins don't talk. Or, if they do, why Richard Dawkins would have hung up on Wilber. Ken Wilber still wants to believe in mystic "goo." He wants a cosmic feel good story, even if his flowery description of the same doesn't have even an ounce of scientific credibility. How do you think Dawkins would respond to this from Wilber [in "On the Nature of Involutionary Givens"]:

"Here is a myth that is sometimes useful in suggesting notions that cannot be grasped dualistically or conceptually in any event: As Spirit throws itself outward (that's called involution) to create this particular universe with this particular Big Bang, it leaves traces or echoes of its Kosmic exhalation. These traces constitute little in the way of actual contents or forms or entities or levels, but rather a vast morphogenetic field that exerts a gentle pull (or Agape) toward higher, wider, deeper occasions, a pull that shows up in manifest or actual occasions as the Eros in the agency of all holons. (We can think of this "pull" as the pull of all things back to Spirit; Whitehead called it "love" as "the gentle persuasion of God" toward unity; this love reaching down from the higher to the lower is called Agape, and when reaching up from the lower to the higher is called Eros: two sides of the same pull). This vast morphogenetic pull connects the potentials of the lowest holons (materially asleep) with the potentials of the highest (spiritually awakened). The involutionary given of this morphogenetic field is a gradient of potentials, not actuals, so that Agape works throughout the universe as a love of gentle persuasion, pulling the lower manifest forms of spirit toward higher manifest forms of spirit--a potential gradient that humans, once they emerged, would often conceptualize as matter to body to

60

mind to soul to spirit. "Spirit" (capital "S"), of course, was (and is) the ever-present ground of all of those manifest waves, equally and fully present in each, but "spirit" (small "s") is also a general stage or wave of evolution: spirit is the transpersonal stage(s) at which Spirit as ground can be permanently realized.

The residue of this involutionary outpouring are various involutionary givens (or items that are given or deposited by involution, items that therefore pre-existed the big Bang and thus are already operating from the moment of the Big Bang forward), the most general of which is the great morphic field of evolutionary potential, a gentle gradient of persuasion pulling all manifest holons back to their ever-present Ground as Spirit--a Kosmic field of Agape, gently pulling evolution into greater and greater consciousness, embrace, inclusion. The universe, it appears, is tilted, and its entire contents are slowly sliding into the Source and Suchness of the entire display. This tilt, this grain to the Kosmos, this Agape, this vast morphogenetic potential, exerts a tender pull on evolution to unfold in waves of greater complexity, greater inclusiveness, greater depth, until the entire Kosmos is included in a prehensive unification that can swallow the Pacific Ocean in a single gulp, hold Mount Everest in the palm of its hand, blink and bring nightfall to the entire universe, smile and bring forth the sun to shine on all creatures great and small."

My hunch tells me that after listening to Wilber's fantasy, Richard Dawkins might just paraphrase Bill Murray's charcter from the original Ghostbusters movie, and exclaim "I have just been slimed by a huge ball of New Age goo."

Wilber wants us to still believe in fairy tales, even if dressed up in pseudo-scientific jargon.

Dawkins on the other hand wants us to finally grow up and admit what we have secretly thought for a long time. We were wrong about the gods and we were wrong about ourselves.

Yes, mysticism can indeed be studied scientifically, but I think we might be a little shocked to see how different it looks when we take off its mythic garb. Even if Wilber wants to redress the old mistress in new tassels, the odd fact remains that mysticism without mystery may kill its original attraction.

Perhaps the most problematic issue confronting transpersonalists is the veracity of inner experiences. For many involved in new religious groups mystical encounters, like near-death and out-of-body excursions, offer evidence of their respective guru's rightful position or succession. This has been especially acute in several Sant Mat related groups, particularly Kirpal Singh's Ruhani Satsang, where mastership disputes are often resolved by resorting to one's inner meditation experiences. But there is a rub in all this that for the most part lies uninspected by those newly initiated.

No doubt a religious devotee may use such experiences as proof for the authenticity of his/her guru or group, but what he/she fails to realize is that there are thousands, if not millions, of people who also claim personal revelations which convince them of the truthfulness of their chosen path. Even Elvis has hundreds of devotees who reported seeing his radiant form at the end of a long dark tunnel when they underwent a near-death experience. So if someone in Memphis can see Elvis in their meditation, are we then supposed to believe in the spiritual mastership of Elvis? Don't get me wrong, I am the first to admit that the King had some great songs during his career, but just because a crew of devoted fans have glimpses of him in the alleged after-life does not constitute documented proof of his spiritual attainment.

I have met scores of New Agers, each initiated by some great guru, who claim to have extraordinary experiences. So what? People can be deceived (like, for instance, Arran Stephens who admitted that he was duped on the inner regions by his experiences with Sant Ajaib Singh). So this issue of inner experiences as proof of a guru's status raises a very important epistemological question: how do we know that what we perceive in mystical practices is truthful or accurate? Now we may come up with any host of supporting evidences, but the fact remains that what one experiences individually in the privacy of meditation is circumscribed by exactly that same feature: *private, personal experience*. What we convey in writing, or what we convey on the telephone, or what we convey by conversation face to face is not evidence of our inner experiences on the spiritual planes, but merely testimony that one can either believe or disbelieve. The naive seeker may accept or reject it as suggestive of truth, but such testimony in itself adds zip to the question of empirical confirmation. Look at the initiates of Thakar Singh or John-Roger Hinkins, each of whom have the same story to tell and it is precisely like their rival counterparts: "I had a mystical experience which convinced me beyond a reasonable doubt that my guru is genuine." The net result is not some universal mystical agreement ("Yes, we do agree that Elvis is the transcendental King"), but rather a plethora of competing accounts, each which patently contradict the other.

What is the primary difference between a fundamentalist Christian and a mystically inclined yogi, especially when it comes to evaluating their ultimate truth claims? Both think they have uncovered the truth: the former by the revealing "Word" of the Bible; the latter by the manifesting inner "Word" of the higher regions. Yet, in both cases, the neophyte is subject to doubt, to skepticism, to deception, since revelations of truth (both inner and outer) are manifold. The Muslims have their *Koran*; the Sikhs have their *Guru Granth Sahib*; and the Christians have their *Bible*. And, for the mystics, yogis, and sages who turn inward what do we find? The Hare Krsnas' see Lord Krishna; the Saivites see Lord Shiva; and Ruhani Satsangis (depending upon your affiliation) see Sant Rajinder Singh, or Sant Baljit Singh, or Sant Sadhu Ram

But, as the argument goes, the devoted mystic will say that his or her experiences are authentic (because of the utter certainty of the encounter) and the experiences of others, especially if they belong to a rival group which splintered off after a succession dispute, are misguided, secondary, or illusory. So what we actually have in effect here in in terms of truth claims is not essentially different than that of a fundamentalist. The mystic is right by virtue of his/her inner attainment and everybody else is wrong (no matter how politely we may gloss over it: karma or chance?) because he/she happened to get the right guru and the right path (and by right we mean "highest").

But notice how the mystic is not calling into question or doubt his/her own truth claims. For example, one rarely finds a completely agnostic posture among disciples about the relative status of his/her guru. Why not? Because just like the fundamentalist he or she is not trained to severely doubt interior revelations of truth, primarily because they appear so real when they occur. It is one thing to state that my inner experiences have convinced me that I am on the right track; quite another to then make judgments on the veracity of other meditators' experiences. Yet even here the person most vulnerable to deception is our self. Paradoxically, the most certain and overwhelming an inner experience appears to be the less likely we are to look for more mundane causations. In other words, the very topography of a mystical encounter tends to blind one from looking for alternative explanations for its originations. Here the Indian word Maya is, surprisingly, apt: "that which betrays its real origin."

To strike a sociological note, it becomes fairly apparent that culture plays a significant role in the ultimate interpretations of inner experiences. What at first glance appears to be a simple, sweet path to enlightenment, turns out to be on closer inspection a political contest over religious claims--claims, I should add, that have been transformed by the cultural landscape of when and where they take place. We may wish that mysticism was devoid of culture, or personal bias, or religious prejudice, but it is almost wholly entrenched in it.

Why? Because we never apprehend inner lights and sounds and beings divorced of their interpretative network. In other words, our socially conditioned minds are always flavoring, always transforming, and always contextualizing whatever we perceive, whether those sights are inner or outer. And it is exactly when my experiences are personal and internal that I am most subject to error. Why? Because we have yet to discern a normative corrective for mystical encounters. To be sure we have templates to gauge inner experiences, their relative efficacy and so on, but since most individuals have no mastery of experiencing OBE's and NDE's we are subject to tremendous imprecision and tremendous speculation. Yet do we admit to this impasse? Do we acknowledge our immaturity in the so-called spiritual arena?

There is something fundamentally skewed when religious converts (of any persuasion and of any methodological bent) begin to believe that they have cornered the market on truth. As one wise saying puts it, "If there really is a God, He/She may find atheism to be less of an insult than religion." The point is obvious: what we know the least about is the very thing we make absolute statements on. Strange, but true. Take Jesus Christ, as a prime, if controversial, example. What do we really know about him? Not very much. Depending on your perspective and the sources that you cite, Jesus emerges as the only begotten Son of God, a Jewish mystic with Gnostic leanings, or a clever, but ultimately misguided magician. The only thing that is absolutely certain about Jesus, at least historically speaking, is that we know less about him than we think. Indeed, the real truth about Jesus' existence is forever buried in the recesses of time.

And yet we have some two billion plus people on this planet right now who more or less believe that if you don't accept the truth claims of Jesus Christ (as provided in certain books that themselves are the result of questionable political processes) you will end up in eternal hell. All of this and we still don't know what he even looked like and what he did for some fifteen years in his teens and early twenties? Couple this with the contradictory and entirely insufficient biographical details contained in the gospels which are the major sources for Jesus' life and you wonder how a Christian can be so assured in their faith. Put bluntly, you wouldn't allow your son or daughter to marry a prospective suitor if the only information you had on them was equal to what we know about Jesus. But there are millions of us who seriously think that we have to make a lifetime, nay eternal, commitment to a person we have never met and know less about than our next-door neighbor.

When it comes to religion and its claims, whether they are based on revealed texts or interior visions, the one common denominator is that we somehow have to check our brains in at the door before entering into the tabernacle of ultimate truth. Yet it is exactly that brain, that three pounds of wonder tissue or glorious meat, as Patricia Churchland so succinctly puts it, that has allowed us to ponder life's ultimate questions. It is that very brain which has led us to pray, to read, to meditate. It is also that very brain which can misinterpret exterior stimuli as well as internal neural firing. My hunch is that before we make any ultimate claims for truth, we understand that we are constantly subject to error.

So the mystic may potentially be better off than the mere believer, who only reads but never actually engages in technical spiritual practices, because he or she gets firsthand experiences of alternate realms of consciousness not merely menu descriptions of them. But this does not mean that the mystic has experienced the "truth" in all its purity and that the mystic somehow "knows" the efficacy of other spiritual teachers or paths. No, what the mystic does in fact know is rather quite simple: a different state of consciousness which he or she interprets according to his/her cultural or religious background. On that score, I do think that mystics are on the right track; it is better to experiment than simply speculate. Yet, the results of those experiments are subject to numerous interpretations, some of which may be better than others. Since we are still at such a preliminary level in our investigation of states of consciousness beyond the waking-rational level, it seems to me to be a much wiser course for us to adopt a stance of honest humility and openness than succumb prematurely to absolute statements or theorizing which in the end cause much more harm than good.

We may want to believe that our chosen (or, in most cases, assigned) religion is the only true path, or that our personal mystical encounters are reflections of universal truth, but when we do so we are only revealing how exquisitely ignorant we really are. It seems to me that the more we acknowledge that exquisite ignorance, instead of suppressing or outright denying it, the better off we will be. And just maybe, like our wise travelers before us (Socrates, Lao Tzu, Nicholas of Cusa), we will realize that learned ignorance is the beginning of wisdom and the cornerstone of truth.

NOTE

If I might interject a personal note, being taught in Catholic schools for some twelve years and teaching religion in their high schools for another five has its own peculiar advantages, especially when it comes to the topic of assessing leaders like Jesus Christ. For one, I was brought up with a clear, univocal, and dogmatic interpretation of his life teachings. No confusion, no room for debate--in sum, Jesus was the Son of God, the axis point of human history, and the ultimate meaning of the universe. The only drawback to a strict Catholic education, though, is that when you begin to ask critical questions about the origins of your religion, you run into deep trouble. I remember vividly my first run-in. When I asked Father Costello, my freshman religion instructor, if Tibetan Buddhists could go to heaven, he unhesitatingly replied, "No, only baptized Christians can enter the Kingdom." Although I was only fourteen years old at the time, I just couldn't swallow the good Father's answer or the convoluted logic he invoked to support it. "You mean to suggest that God plays geographical favorites?" Or so my reasoning went, but to no avail. I ended up getting reprimanded in front of the class for being out of line and disrespectful. Needless to say, my doubts about the efficacy of Catholic dogma grew exponentially after this incident.

6 | The Disneyland of Consciousness

Dr. Don Salmon's recent essay, *Shaving Science with Ockham's Razor*, was a delight to read. I particularly appreciated how he organized his ideas and how he displayed a disarming parenthetical humility (such as when he checks his musings with "though it may be so"). I am looking forward to how he will develop a "methodology to detect the presence of consciousness in the world" which he purports will be the "the subject of a future video series."

One of the linchpins in Dr. Salmon's argument is that "there is no scientific finding that compels us to think of 'X' as either conscious or non-conscious, living or non-living, intelligent or non-intelligent." According to his argument this is "because scientific methods do not tell us what lies beyond our mind-constructed precepts and constructs." Dr. Salmon's utilizes a number of quotes touching upon our inability to "objectify" reality when we aim to scientifically understand it through our limited perceptions. In this regard, he backs up his assertions with three famously pregnant quotes from quantum theorists:

"[In the study of modern physics] we can never understand what events are, but must limit ourselves to describing the patterns of events in mathematical terms; no other aim is possible. Physicists who are trying to understand nature may work in many different fields and by many different methods; one may dig, one may sow, one may reap. But the final harvest will always be a sheaf of mathematical formulae. These will never describe nature itself. . . . [Thus] our studies can never put us into contact with reality."
--*Sir James Jeans*

"What we observe is not Nature in itself but Nature exposed to our method of questioning."
--*Werner Heisenberg*

"Human beings are stuck in a Midas-like predicament: we can't directly experience the true texture of reality because everything we touch turns to matter."

--Nick Herbert

I think Niels Bohr and others of like ilk would have no problem with Dr. Salmon's assertions since the Copenhagen interpretation of quantum mechanics essentially says, "It is wrong to think that the task of physics is to find out how Nature is. Physics concerns what we say about Nature."

In addition, neurologists and neuroscientists are keenly aware that everything we experience is modulated through our brains. This is perversely summarized in the shorthand quip, "Try thinking of something without your brain."

Salmon goes a step further and focuses on our a priori experience of consciousness first. He very straightforwardly states, "All that we experience is within consciousness."

This is, of course, so fundamentally obvious that we sometimes overlook or neglect what such a radical supposition implies. John Searle, the well-known philosopher and linguist from U.C. Berkeley, has long been a champion of a fuller explanation of consciousness, one that doesn't ignore the apparent indissoluble subjective experience that is at the heart of our "I" awareness. As Searle explains, "The very fact of subjectivity, which we were trying to observe, makes such an observation impossible. Why? Because where conscious subjectivity is concerned, there is no distinction between the observed and the thing observed. . . . Any introspection I have of my own conscious state is itself that conscious state."

Searle argues for the irreducibility of consciousness, but ironically adds the odd caveat that it has "no deep consequences."

If I understand Dr. Salmon's essay correctly, he would have a qualm with Searle's categorical dismissal, especially in light of what a first person understanding can bring to bear on "reality".

Searle is not in isolated company, as David Chalmers, author of the widely cited book *The Conscious Mind*, also champions consciousness as a fundamental feature in the universe, and not merely a secondary byproduct.

Interestingly, though, Searle and Chalmers dramatically part company with each other's over what such ontological primacy means.

Francisco J. Varela and Jonathan Shear critique both philosophers because their "adherence to the pertinence of first-person experience [has not been] followed with methodological advances." Additionally, they point out, "The mental thus does not have any sound manner to investigate itself, and we are left with a logical conclusion, but in a pragmatic and methodological limbo. This is not unlike the limbo in Ray Jackendoff's views, where in his own manner he also claims the irreducibility of consciousness, but when it comes to method is tellingly silent. He does claim that insights into experience act as constraints for a computational theory of mind, but follows with no methodological recommendations except 'the hope that the disagreements about phenomenology can be settled in an atmosphere of mutual trust' (Jackendoff, 1987, p. 275)."

Varela and Shear want to move beyond the impasse that seems to have confronted other philosophers of the mind. In this regard, they suggest that there is a necessary "circulation" between first (subjective) person and third (objective) person accounts. They argue,

"Setting the question as we just did, the next point to raise is what is the status of first-person accounts? In some basic sense, the answer cannot be given a priori, and it can only unfold from actually exploring this realm of phenomena, as is the case in the contributions presented herein. However let us state at the outset some thorny issues, in order to avoid some recurrent misunderstandings.

First, exploring first-person accounts is not the same as claiming that first-person accounts have some kind of privileged access to experience. No presumption of anything incorrigible, final, easy or apodictic about subjective phenomena needs to be made here, and to assume otherwise is to confuse the immediate character of the givenness of subjective phenomena with their mode of constitution and evaluation. Much wasted ink could have been saved by distinguishing the irreducibility of first-person descriptions from their epistemic status."

Second, a crucial point in this Special Issue has been to underline the need to overcome the 'just-take-a-look' attitude in regards to experience. The apparent familiarity we have with subjective life must give way in favour of the careful examination of what it is that we can and cannot have access to, and how this distinction is not rigid but variable. It is here that methodology appears as crucial: without a sustained examination we actually do not produce phenomenal descriptions that are rich and subtly interconnected enough in comparison to third-person accounts. The main question is: How do you actually do it? Is there evidence that it can be done? If so, with what results?

Third, it would be futile to stay with first-person descriptions in isolation. We need to harmonize and constrain them by building the appropriate links with third-person studies. In other words we are not concerned with yet another debate about the philosophical controversies surrounding the first-person/third-person split, (a large body of literature notwithstanding). To make this possible we seek methodologies that can provide an open link to objective, empirically based description. (This often implies an intermediate mediation, a second-person position, as we shall discuss below.) The overall results should be to move towards an integrated or global perspective on mind where neither experience nor external mechanisms have the final word. The global perspective requires therefore the explicit establishment of mutual constraints, a reciprocal influence and determination (Varela, 1996).

In brief our stance in regards to first-person methodologies is this: don't leave home without it, but do not forget to bring along third-person accounts as well. This down-to-earth pragmatic approach gives the tone to the contributions that follow. On the whole, what emerges from this material is that, in spite of all kinds of received ideas, repeated unreflectingly in recent literature of philosophy of mind and cognitive science, first-person methods are available and can be fruitfully brought to bear on a science of consciousness. The proof of the pudding is not in a priori arguments, but in actually pointing to explicit examples of practical knowledge, in case studies.

I am not sure whether or not Dr. Salmon would agree wholly or partially or not at all with Varela's and Shear's positional methods, but I do think that these discussions are worthy of merit and that cross volleys from differing directions on this subject are helpful, even if we might on occasion disagree with their intended import.

I particularly appreciated how Dr. Salmon invoked (with obvious approval) William James' approach to the subject of consciousness. Ironically, the uber champion of mind to brain reductionism, the late Francis Crick, also found much to admire in James' approach to self-reflective awareness, as have other working neurobiologists from Gerald Edelman to V.S. Ramachandran.

In this context, I thought it might be useful and perhaps of some interest if we explored how we recognize consciousness in others by focusing on examples of where individuals have mistakenly imputed human awareness upon inanimate objects only to be flabbergasted after discovering their mistaken conflations. My personal hunch is that such a pathway may give us some indications how a future science of consciousness may proceed.

GREAT MOMENTS WITH MR. LINCOLN

In the late 1960s one of the more popular attractions at Disneyland was *Great Moments with Mr. Lincoln*, which displayed one of the first audio-animatronic devices that featured President Abraham Lincoln. It was impressively life-like at the time, even though the machinery was still in a rudimentary stage.

I will never forget an incident that transpired just after the show was completed. An older lady, presumably in her 80s, had mistakenly believed that the audio-animatronic was a real human being. Indeed, she thought a very talented actor performed it. She was so impressed by the performance that she went up to the usher to see if she could get the performer's autograph. You can imagine her shock when she was informed that it was not a human being but a machine.

PIRATES OF THE CARIBBEAN

Coincidentally, the same thing happened to me when I went on the Pirates of the Caribbean when it first opened up. As we were navigating through the underground tunnels with an assortment of decorated pirates regaling their

73

exploits, I looked on one pirate overhanging from the bridge who appeared to look me right in the eye. I turned towards my friend and said, "That pirate is a real person." Indeed, I was so convinced that the apparent audio-animatronic was an actor dressed in costume that I refused to believe otherwise until several years later when I realized that it was the lighting that had played a trick on my eyes.

THE HAUNTED MANSION

Although it was initially billed as a scary ride, the *Haunted Mansion* turned out to be quite tame. However, there was one segment in the ride where a friend of mine got the creeps, since he assumed that the head (sans a body) within the crystal ball was a real person. He hadn't realized how far holographic technology had come at that point and couldn't imagine that it was merely a vaporous projection.

I bring up these three examples because they underline something fundamental in our assessment of the consciousness of others. We can be easily duped. Not only can we impute conscious intentionality onto machine operated mannequins that lack it, we can even do it photographic film.

Yet phenomenologically speaking, our own experience at the time of interacting with an audio-animatronics seems essentially the same as when we talk to certain humanoids. In other words, that which we believe is conscious turns out on closer inspection to be unconscious, at least in the commonsense ways that we use such terms in our day to day lives.

But we don't need to go Disneyland to discover this, since we already have firsthand experience of innumerable conflations when we fall asleep and dream.

In a strong dream, so many characters come alive and we interact as if each of them is real. Only when we wake up do we acknowledge that everything that occurred in the dream was simulated by us. We are, in sum, dreaming ourselves in various guises, even if we may be deceptively tricked to believe otherwise. Such is the confusing nature of our own self-awareness that we even objectify our own personas in various garbs and believe them to be ontologically apart from our own neural projections.

Consciousness is a fantastic virtual simulator and because of its inclusivity and insular engineering it has an inherent tendency to believe its own machinations as exterior to itself and not as the byproduct of its own interiority.

This is one strange loop, as Douglas Richard Hofstadter brilliantly opined in his book of the same title, *I am a Strange Loop* and in his earlier and now classic tome, *Gödel, Escher, Bach: An Eternal Golden Braid*.

The problem of "other" minds has long been a philosophical conundrum among thinkers from centuries past. The difficulty gets even stickier when we pause and realize that we don't even have a full and complete access to our own minds and its labyrinth like permutations since there are so many processes within our own skull and body which we remain dutifully unaware.

Saying that all things are experienced through consciousness unnecessarily reifies the very word consciousness since awareness is not a "thingy" but a process as fluid as the tons of water cascading down Niagara Falls.

THE SILVER MIME

Not only can we confuse a robot for a human being, we can also do the reverse. I remember when my son Shaun was very young and we were walking through a shopping district and we saw a mime painted and dressed entirely in silver. He didn't move or twitch or even bat an eye lash. He was utterly frozen still, or so it seemed. My son, who was barely two or three at the time, thought it was some sort of robot and when I said it was a human being he simply refused to believe me. When the mime finally did emerge from his robotic trance, his movements were so geometrically precise that Shaun left convinced that the Silver Man was just the latest toy to come onto the market.

Why is any of this pertinent in our discussions on consciousness and what it means? The answer is simple:

determining whether someone is conscious or unconscious is, as far as we can tell at this stage, the result of how we individually interpret a set of behavioral actions. Moreover, consciousness is not merely one thing, but rather a series of movements and reflections and moments turning this way and that in a constant parade of attentional space.

MULTIPLE PERSONALITIES

In the 1990s I had a student who was diagnosed with 31 different personalities. Certain school officials asked me if I wanted her removed from the classroom since it was an unpredictable situation and had already caused discomfort for other professors and students. I declined the request since I found Becky (not her real name) to be a remarkably bright and gifted young woman, despite the fact that she would assume several different personalities during the day.

Whenever Becky attended my class, she dressed up as if she was a stuffy librarian conservatively dressed, with her hair tied up in a bun, and quite prim and proper. But an hour or so after my class, whenever she visited me in my office she dressed down as if she was a surfer hippie from the late 1960s. Occasionally she would dress up as a man and act as if she held a day job as a plumber.

I never quite knew what personality she would display on any given day, except that I told myself to remain calm and just respond normally regardless of appearances. Eventually, Becky and I became good friends and I discovered what had happened that generated such a kaleidoscope of personas. When Becky was very young she was abused by her stepfather who broke her back when she was just six years old. He subsequently stuck her in a dark closet for a year and a half. Apparently, all of this abuse was justified by her stepfather as part of his newfangled religious beliefs which centered on some peculiar and dogmatic interpretations of the Bible, particularly the Old Testament. Becky's way of coping with such a horrific upbringing was to dissociate different parts of herself into distinct personas, so much so that they became an inherent part of how she dealt with the world.

Unlike the vast majority of human beings who also have multiple personas or selves (as Errol Flynn, the famous movie actor of the 1930s and 1940s tellingly wrote, "I am an octagon of contradictions which in itself may be no contradiction"), but which are mostly transparent to each other, Becky had developed—consciously or unconsciously—separate walls between each of her personalities and because of this there was a jolting disconnect for those who came into contact with her.

Who was the real Becky one might venture to ask? Well, the question itself is misleading and I would suggest is unanswerable, since it assumes an agreed upon ontology about reality that (at least at this level) is unknown to us.

We are it seems in an intractable position when it comes to answering ultimate questions and as such we are forced to take a more pragmatic view of things. William James, writing a century ago, captured this when he wrote,

"The pragmatic method is primarily a method of settling metaphysical disputes that otherwise might be interminable. Is the world one or many? – fated or free? – material or spiritual? – here are notions either of which may or may not hold good of the world; and disputes over such notions are unending. The pragmatic method in such cases is to try to interpret each notion by tracing its respective practical consequences. What difference would it practically make to any one if this notion rather than that notion were true? If no practical difference whatever can be traced, then the alternatives mean practically the same thing, and all dispute is idle. Whenever a dispute is serious, we ought to be able to show some practical difference that must follow from one side or the other's being right."

William James, touching upon a procedural method versus a final concluding prolegomena continues, "Metaphysics has usually followed a very primitive kind of quest. You know how men have always hankered after unlawful magic, and you know what a great part in magic words have always played. If you have his name, or the formula of incantation that binds him, you can control the spirit, genie, afrite, or whatever the power may be. Solomon knew the names of all the spirits, and having their names, he held them subject to his will. So the universe has always appeared to the natural mind as a kind of enigma, of which the key must be sought in the shape of some illuminating or power-bringing word or name. That word names the universe's principle, and to possess it is after a fashion to possess the universe itself. 'God', 'Matter', 'Reason', 'the Absolute', 'Energy', are so many solving names. You can rest when you have them. You are at the end of your metaphysical quest. But if you follow the pragmatic method, you cannot look on any such word as closing your quest. You must bring out of each word its practical cash-value, set it at work within the stream of your experience. It appears less as a solution, then, than as a program for more work, and more particularly as an indication of the ways in which existing realities may be changed. Theories thus become instruments, not answers to enigmas, in which we can rest. We don't lie back upon them, we move forward, and, on occasion, make nature over again by their aid. Pragmatism unstiffens all our theories, limbers them up and sets each one at work."

William James nicely summarizes how such a practical view of things would proceed, "No particular results then, so far, but only an attitude of orientation, is what the pragmatic method means. The attitude of looking away from first things, principles, 'categories,' supposed necessities; and of looking towards last things, fruits, consequences, fasts."

In light of Becky's many personas and in light of our discussion of consciousness, it becomes obvious that we are stuck to the world of appearances, even the appearance of our own deeply felt self-awareness. Thus, although if we may heartily disagree with Francis Crick's radical reductionism of consciousness as merely a set of bundled neurons, we can readily see how his agenda can proffer undiscovered vistas that would otherwise be lost in a purely consciousness first approach. In other words, regardless of the ontology of what consciousness ultimately is, we can make progress by analyzing what it "appears" to be and how such appearances may be modified by food, drink, drugs, sex, meditation, and so forth. This doesn't do away with first-person explanations at all, since what we feel or experience isn't necessarily divorced from what we image. No dentist worth his reputation discounts the reportage of his patient's first-hand experiences of pain.

A doctor, for instance, may not precisely know what his or her patient is undergoing (even if she had undergone a similar procedure herself) when she performs surgery on them, but by keeping a keen eye and ear on her patient (and also monitoring internal machinations such as blood pressure and the like), the doctor can discover patterning that is correlative to the patient's internal experiences.

The hard problem in studying consciousness is oftentimes referred to as the "qualia" referendum (e.g., how can we possibly know what the subjective experience of "redness" is for another person?). It is difficult enough to even describe our own inner (and seemingly unique) subjective experiences, much less properly analyze what another person may or may not be feeling or seeing or experiencing.

Is consciousness a Skinnerian black box, whereby we are pushed to the hinterlands of behaviorism 101 just to monitor external responses to given stimuli?

Or, can we by our own self-conscious understandings somehow bridge the gap between first and third person narratives? Can we, in sum, develop a communicative system where I meets Thee?--A sort of Martin Buber form of consciousness studies?

These are the kinds of questions that Dr. Solman's essay brings forth and the more practical neuroscientists working in the field today have tended to focus their research efforts on parts of the great puzzle, hoping that biting off smaller chunks the larger prey—a consciousness holistically understood—will one day become an achievable goal.

Others like Colin McGinn, the British philosopher now at the University of Miami, have argued that consciousness is an inherent mystery that is essentially non solvable. This so-called new mysterianism is explained thusly,

"New mysterians argue that their belief that the hard problem is unresolvable is not a presupposition, but is a logical conclusion reached by thinking carefully about the

79

issue. The standard argument is as follows: Subjective experiences by their very nature cannot be shared or compared. Therefore it is impossible to know what subjective experiences a system (other than ourselves) is having. This will always be the case, no matter what clever scientific tests we invent. Therefore, although a person may know that they have qualia, they cannot meaningfully discuss these qualia from a third-person point of view, and the topic will remain mysterious and unresolvable."

Patricia and Paul Churchland, the eminent neurophilosphers from UCSD, have little tolerance for this approach and have long argued that a robust eliminative materialism, grounded in the latest discoveries of neuroscience, can indeed unravel the mystery of consciousness itself. But in order to do such old and outdated (and misleading) folk psychological terminology has to be replaced with a more accurate neurological language. Thus, the very word consciousness has to be redefined to more accurately reflect what we ourselves experience moment to moment. It might be useful as an all-purpose touch phrase in polite conversation, but its very vagueness creates what may turn out to be an illusory and dualistic reification. For example, an untrained ear may listen to a 40 member orchestra, with a wide array of instruments, and say after the playing that it was a wonderful "song" implying a singularity of sorts since they didn't differentiate the various accompanying sounds (a flute, a piano, a violin) each of which served combinatorially to produce the desired effect. But a trained ear would realize that there were so many different sounds being produced and that the beautiful harmony within the song was the result of many varying features and that the effect of unification was itself the illusory byproduct of divergence.

Could our own "training" in self-reflection lead to similar results, whereby we unpack our perceived unity in consciousness and discover anew that it is anything but? Could an evolving neuroscience, and its attendant eliminative materialist language, lead us to a refashioning and a radical revising of how we talk about ourselves? Will the very term consciousness give way just as phlogiston and demonic possession have in the past once we have better understood the physical causes underlying such apparent phenomena?

The jury is still out on these and other questions and we will have to wait further deliberations until we can be satisfied with a final verdict.

However, there has been some remarkable progress in creating artificial systems that give at least the appearance of intentionality, intelligence, and consciousness. Ironically, by focusing on making artificially intelligent machines we may better understand what forms the "apparent" basis of self-conscious navigating systems. Just as Francis Crick and James Watson discovered the double helix structure of DNA (and the secret to genetic coding) by their endless model building (none of which literally used organic material, but was rather done tinker toy like using metals), perhaps by artificially building models of what appear life-like or conscious we may get a deeper grasp of our own awareness.

From an absolutist perspective, I can appreciate Dr. Salmon's claim that "there is no scientific finding that compels us to think of 'X' as either conscious or non-conscious, living or non-living, intelligent or non-intelligent," but the truth is that we clearly have very practical ways of going about determining such demarcations.

In fact, our very survival is predicated upon making day-to-day decisions over these very issues. Whenever I surf I tend not to confuse a surfboard with a surfer, neither do I confuse a mound of seaweed with a shark. When I was younger and attended a high school dance in our local gym, I didn't confuse a beautiful young girl with the basketball hoop. Moreover, whenever a policeman has stopped me in my car I am very clear in not conflating him with a rail guarding. We make such distinctions all the time, even if we don't have deep and persuasive philosophical reasons justifying why. Yet, do I seriously think that our decision-making processes are outside of the realm of science? No, not at all, since we can easily identify a whole set of external markers which help us make such split second decisions about whether something is living or not or whether something is conscious or not.

Our practical nature in this regard (played out in a real and observable empirical arena) is obviously amenable to an objectivist and scientific study, even if we cannot in a Kantian way get at the "thing in itself."

Or, perhaps following Nietzsche's line of reasoning, there may be no "essence" to speak of, no "thing itself" as such, and thus our very language use betrays the subject/object dualism that we are heroically trying to resolve.

In this regard, one could better understand why a practical approach to the study of consciousness may be the most fruitful one in the long run and that resorting to tautological syllogisms at the start ends up, paradoxically, to be a nonstarter.

THE TURING DILEMMA

Alan Turing raised the stakes on what constitutes intelligence when he wrote "Computing Machinery and Intelligence" in 1950 in the journal *Mind* about whether or not machines could think based on what he termed an "imitation game." Turing explains, "It is played with three people, a man (A), a woman (B), and an interrogator (C) who may be of either sex. The interrogator stays in a room apart from the other two. The object of the game for the interrogator is to determine which of the other two is the man and which is the woman. He knows them by labels X and Y, and at the end of the game he says either 'X is A and Y is B' or 'X is B and Y is A.' The interrogator is allowed to put questions to A and B thus: C: Will X please tell me the length of his or her hair? Now suppose X is actually A, then A must answer. It is A's object in the game to try and cause C to make the wrong identification. His answer might therefore be: 'My hair is shingled, and the longest strands are about nine inches long.' In order that tones of voice may not help the interrogator the answers should be written, or better still, typewritten. The ideal arrangement is to have a teleprinter communicating between the two rooms. Alternatively the question and answers can be repeated by an intermediary. The object of the game for the third player (B) is to help the interrogator. The best strategy for her is probably to give truthful answers. She can add such things as 'I am the woman, don't listen to him!' to her answers, but it will avail nothing as the man can make similar remarks. We now ask the question, 'What will happen when a machine takes the part of A in this game?' Will the interrogator decide wrongly as often when the game is played like this as he does when the game is played between a man and a woman? These questions replace our original, 'Can machines think?'"

Later, Turing's imitation game was modified by a number of computer scientists and today it has evolved into several different versions. There is even an annual artificial intelligence competition held in Alan Turing's honor. The winner is awarded the *Loebner Gold Medal*. As the official website explains, "The Loebner Prize for artificial intelligence (AI) is the first formal instantiation of a Turing Test. The test is named after Alan Turing the brilliant British mathematician. Among his many accomplishments was basic research in computing science. In 1950, in the article Computing Machinery and Intelligence which appeared in the philosophy journal *Mind*, Alan Turing asked the question "Can a Machine Think?" He answered in the affirmative, but a central question was: "If a computer could think, how could we tell?" Turing's suggestion was, that if the responses from the computer were indistinguishable from that of a human, the computer could be said to be thinking. This field is generally known as natural language processing. In 1990 Hugh Loebner agreed with The Cambridge Center for Behavioral Studies to underwrite a contest designed to implement the Turing Test. Dr. Loebner pledged a Grand Prize of $100,000 and a Gold Medal (pictured above) for the first computer whose responses were indistinguishable from a human's. Such a computer can be said 'to think.' Each year an annual prize of $2000 and a bronze medal is awarded to the most human-like computer. The winner of the annual contest is the best entry relative to other entries that year, irrespective of how good it is in an absolute sense."

Not unexpectedly, given the rapid progress in technology, the Loebner Prize winners have continually gotten better at appearing "human" like.

What this test and others like it suggest is that given enough time we should be able to develop artificial devices of sufficient complexity which will consistently trick us into believing that they are human and not robotic. But this very trickery should give us a clue on how to better understand the appearance of self-awareness both to ourselves and to others.

And maybe in the not so distant future time we will be unable (or unwilling or unmotivated?) to differentiate an artificially engineered intelligent life form (replete with a convincing self-conscious display system) from a biological humanoid. Already it is becoming increasingly difficult for some to distinguish a computerized voice from a human one and given enough time and enough computational power everything a human being does so naturally may be easily mimicked. We don't have to accept every detail of Ray Kurzweil's controversial "Singularity" hypothesis to acknowledge that CGI is becoming so realistic as to become a "reality" without seams (where what is real blurs with what is artificial).

This was explicitly driven home to me when I showed a clip of a surfer riding a gigantic wave which was entitled "tsunami surfing." The majority of my students thought it was the result of computer-generated imagery. However, much to their amazement, I explained that it was genuine footage of a real surfer (Mike Parsons) riding a real wave (Jaws) in a real location (Maui).

Conversely, when I played a computerized voice named "Peter" that I had downloaded as a 99 cent app on my Apple iPad, the majority of my students believed that the voice was a human being talking, not the result of a sophisticated computer program.

Dr. Salmon ends his essay with an intriguing thought experiment in which he enjoins his readers as follows:

"If your general view is that X is conscious, spend more time studying the passage while assuming that X is non-conscious. If you think X is non-conscious, challenge yourself, see how much support you can find for the idea that X might involve consciousness or intelligence in some way."

Interestingly, I think the idea that consciousness is a virtual simulator can dovetail nicely with thought experiments of this kind, since "imagining" what another person may consciously intend does indeed have dramatic consequences on how we may or may not react to a given situation. Of course, the opposite is true as well, since if we believe that the object in our purview is unconscious or unintelligent our reactions may differ. But in both cases we are "simulating" a situation within our consciousness and acting accordingly.

Gerald Edelman, the distinguished Nobel laureate, has in a number of his books touched upon two fundamental forms of awareness, what he calls first and second nature. In a recent interview with *Discover* magazine, Edelman explains, in brief, his bipartite view of consciousness,

"There is every indirect indication that a dog is conscious—its anatomy and its nervous system organization are very similar to ours. It sleeps and its eyelids flutter during REM sleep. It acts as if it's conscious, right? But there are two states of consciousness, and the one I call primary consciousness is what animals have. It's the experience of a unitary scene in a period of seconds, at most, which I call the remembered present. If you have primary consciousness right now, your butt is feeling the seat, you're hearing my voice, you're smelling the air. Yet there's no consciousness of consciousness, nor any narrative history of the past or projected future plans. Humans are conscious of being conscious, and our memories, strung together into past and future narratives, use semantics and syntax, a true language. We are the only species with true language, and we have this higher-order consciousness in its greatest form. If you kick a dog, the next time he sees you he may bite you or run away, but he doesn't sit around in the interim plotting to remove your appendage, does he? He can have long-term memory, and he can remember you and run away, but in the interim he's not figuring out, 'How do I get Kruglinski?' because he does not have the tokens of language that would allow him narrative possibility. He does not have consciousness of consciousness like you."

Edelman then goes on to explain why such a "Second" nature may be of evolutionary advantage in human beings, "The evolutionary advantage is quite clear. Consciousness allows you the capacity to plan. Let's take a lioness ready to attack an antelope. She crouches down. She sees the prey. She's forming an image of the size of the prey and its speed, and of course she's planning a jump. Now suppose I have two animals: One, like our lioness, has that thing we call consciousness; the other only gets the signals. It's just about dusk, and all of a sudden the wind shifts and there's a whooshing sound of the sort a tiger might make when moving through the grass, and the conscious animal runs like hell but the other one doesn't. Well, guess why? Because the animal that's conscious has integrated the image of a tiger. The ability to consider alternative images in an explicit way is definitely evolutionarily advantageous."

Going back in time, Edelman suggests how early forms of consciousness may have evolved: "About 250 million years ago, when therapsid reptiles gave rise to birds and mammals, a neuronal structure probably evolved in some animals that allowed for interaction between those parts of the nervous system involved in carrying out perceptual categorization and those carrying out memory. At that point an animal could construct a set of discriminations: qualia. It could create a scene in its own mind and make connections with past scenes. At that point primary consciousness sets in. But that animal has no ability to narrate. It cannot construct a tale using long-term memory, even though long-term memory affects its behavior. Then, much later in hominid evolution, another event occurred: Other neural circuits connected conceptual systems, resulting in true language and higher-order consciousness. We were freed from the remembered present of primary consciousness and could invent all kinds of images, fantasies, and narrative streams."

I think the study of consciousness is an open field of inquiry and, as John Searle once indicated in an interview on this subject, all thinkers on this subject are welcome to have a stab at it.

However, I do believe that there are some approaches (such as the neurobiological one) that even if they turn out to be incomplete or insufficient will undoubtedly produce more fruitful byproducts and offspring than others. In this regard, the more empirical strategies (focusing on the visual cortex or the olfactory nerve or mirror neurons) have been wildly successful in helping us understand combinatorially some of the outstanding features that often play a key feature in what we term consciousness, even if they ultimately come up short. While one may argue with Francis Crick's ultimate take on the biochemical basis of self-awareness, there is no getting around the fact that his practical approach to the subject (following one of William James' more practical routes) has paid off handsomely with regard to better appreciating how neuronal clusters interact, particularly in relation to our notions of vision.

But Crick's pathway shouldn't serve as a roadblock to others who have contrarian views on the subject. We are still in the preliminary stages in our grasp of this most fundamental of subjects and I do believe that Dr. Salmon's perspective (and others of like mind) should be taken seriously, even if it cuts across the prevalent model making within certain neuroscientific circles.

A recent book by David Kaiser, a professor of the history of science at M.I.T., entitled *How the Hippies Saved Physics* provides a compelling argument that a group of rag tag physicists (most without steady jobs and most who held unorthodox ideas) changed the course of physics by following a radical and unconventional path where psychedelic drugs, psychic warfare, and Bell's theorem interfaced. As the publisher W.W. Norton & Company book summary explains,

"For physicists, the 1970s were a time of stagnation. Jobs became scarce, and conformity was encouraged, sometimes stifling exploration of the mysteries of the physical world. Dissatisfied, underemployed, and eternally curious, an eccentric group of physicists in Berkeley, California, banded together to throw off the constraints of the physics mainstream and explore the wilder side of science. Dubbing themselves the 'Fundamental Fysiks Group,' they pursued an audacious, speculative approach to physics. They studied quantum entanglement and Bell's Theorem through the lens of Eastern mysticism and psychic mind-reading, discussing the latest research while lounging in hot tubs. Some even dabbled with LSD to enhance their creativity. Unlikely as it may seem, these iconoclasts spun modern physics in a new direction, forcing mainstream physicists to pay attention to the strange but exciting underpinnings of quantum theory."

Therefore, I think it is wise that even as others and I may promote an intertheoretic and reductionistic approach to the study of consciousness, others with a more holistic viewpoint, such as the one suggested by Dr. Salmon, be given serious and deep consideration. I know for myself that I am looking forward to Dr. Salmon's future offerings, even if my skeptical antenna is pulsing out warning signals.

"The experience of consciousness is a self-referential loop, where our I gets irretrievably intertwined with the world we experience."
--Andrea Lane

7 | Sam Harris' Mobius Strip

Therefore, although science may ultimately show us how to truly maximize human well-being, it may still fail to dispel the fundamental mystery of our mental life. That doesn't leave much scope for conventional religious doctrines, but it does offer a deep foundation (and motivation) for introspection. Many truths about ourselves will be discovered in consciousness directly.
                              --Sam Harris

"But sometimes I think it's just like an on-off switch. Click and you're gone."
                              --Steve Jobs

Is the very notion of a "scientific" study of consciousness oxymoronic from the start? Or, is it just one half of a Mobius-like endeavor that can never succeed unless the fundamental subjectivity inherent in such self-reflective awareness also takes central stage in what appears to be dyadic melody, where being the instrument and listening to the instrument are indissolubly intertwined?

Sam Harris, the first of the four famous horsemen of atheism (his first book, *The End of Faith*, has been credited with sparking the nascent neo-atheist movement now sweeping throughout the USA and abroad), has in a recent two part article ["The Mystery of Consciousness"] strongly argued that consciousness is a mystery that resists an objective scientific explanation.

As Harris writes, "The problem, however, is that no evidence for consciousness exists in the physical world. Physical events are simply mute as to whether it is "like something" to be what they are. The only thing in this universe that attests to the existence of consciousness is consciousness itself; the only clue to subjectivity, as such, is subjectivity. Absolutely nothing about a brain, when surveyed as a physical system, suggests that it is a locus of experience.

Were we not already brimming with consciousness ourselves, we would find no evidence of it in the physical universe—nor would we have any notion of the many experiential states that it gives rise to. The painfulness of pain, for instance, puts in an appearance only in consciousness. And no description of C-fibers or pain-avoiding behavior will bring the subjective reality into view."

Of course Dennett and Churchland (Tufts University and UCSD respectively) have a different take and Harris readily admits this when he writes, "And, again, I should say that philosophers like Daniel Dennett and Paul Churchland just don't buy this. But I do not understand why. My not seeing how consciousness can possibly be an illusion entails my not understanding how they (or anyone else) can think that it might be one. I agree, of course, that we may be profoundly mistaken about consciousness—about how it arises, about its connection to matter, about precisely what we are conscious of and when, etc. But this is not the same as saying that consciousness itself may be entirely illusory. The state of being utterly confused about the nature of consciousness is itself a demonstration of consciousness."

The mystery of consciousness in Harris' purview is very simply, even if ultimately profoundly, obvious to those with this all-pervasive feeling of awareness. We are inside looking outside and when we flip the proceedings to understand the former with the latter we find something elemental missing in the equation. And so we should, since the very dualism of speaking of in/out by necessity brings an impassable confusion to the proceedings. To understand consciousness, Harris reasons, necessitates being conscious and therein lies the difficulty of trying to get an "outside" handle of what in terms of lived through experience has no separable exterior as such.

The first person cannot be translated into a third person descriptive narrative without losing the very essence of what it is to be aware. Yes, Harris admits, that we can learn much about the brain and all sorts of correlative things that occur within our skull when we are conscious, but the qualia doesn't transfer wholly to any sort of reductive explanation.

Harris explains further, "For these reasons, it is difficult to imagine what experimental findings could render the emergence of consciousness comprehensible. This is not to say, however, that our understanding of ourselves won't change in surprising ways through our study of the brain. There seems to be no limit to how a maturing neuroscience might reshape our beliefs about the nature of conscious experience. Are we fully conscious during sleep and merely failing to form memories? Can human minds be duplicated or merged? Is it possible to love your neighbor as yourself? A precise, functional neuroanatomy of our mental states would help to answer such questions—and the answers might well surprise us. And yet, whatever insights arise from correlating mental and physical events, it seems unlikely that one side of the world will be fully reduced to the other."

In a comparative way I have found this same conundrum in surfing. Whenever I sit on the beach and watch the waves at one of my favorite surf haunts, I mind-surf it, imagining that I am bottom turning, lip launching, or getting deeply buried within the tube. Yet, for all the times I have "objectified" being in the ocean and simulating that I am sliding down an almond shaped cylinder, it is doesn't come close to actually being in the water and being immersed into the liquid lines. I have often tried to explain to my two young sons what it is actually like to be inside a hefty tube, but no matter how detailed or refined my tale may be, I know from my own experience that unless they have a tube ride themselves such third person objectifications will not provide the necessary subjective encounter… try as I might.

Describing an experience is, to Harris' view, not the same as being the experience, regardless of how sophisticated our neuroscience becomes. It is for this reason that Sam Harris has championed exploring consciousness via direct introspection. In a rich and pregnant article entitled, "What's the Point of Transcendence?" Harris opines,

"Transcendent experiences, in so far as they are usually temporary, are often surrounded by a penumbra of other states and insights. Just as one can glimpse deeper strata of well-being, and briefly see the world by their logic, one can notice the impediments to feeling this way in each subsequent moment. There is no question that all of these mental states have neurophysiological correlates—but the neurophysiology often has subjective correlates. Understanding the first-person side of the equation is essential for understanding the phenomenon. Everything worth knowing about the human mind, good and bad, is taking place inside the brain. But that doesn't mean that there is nothing to know about the qualitative character of these events. Yes, qualitative character can be misleading, and certain ways of talking about it can manufacture fresh misunderstandings about the mind. But this doesn't mean that we can stop talking about the nature of conscious experience. At one level, there is nothing else to talk about."

To be sure, Harris is not introducing anything new here, except that it is refreshing to read an atheistic neuroscientist giving voice to a millennial old mystical tradition of exploring the source of awareness by focusing on awareness itself.

But what is it that makes the subject of consciousness recalcitrant to explanation, particularly when science has been so successful in explaining such vagaries as the process of genetics and the formation of galaxies?

Perhaps the Mobius strip is a useful, albeit limited, metaphor here to invoke since its unusual properties on first sight boggle the mind: "a surface with only one side and only one boundary component. It has the mathematical property of being non-orientable." Analogously, the difficulty we have with studying consciousness is precisely that we cannot communicate what it is without losing the very quality that makes it such. Imagine that you wanted to convey what it was like to dream to someone who never dreamt and only had access to a waking state. How could you communicate dreaming without dreaming itself? Harris suggests strongly that you cannot and that any attempt to circumvent the obvious is simply unintelligible. How can one convey something that is non-orientable to that which is only orientable, if one has to forego the very thing that would communicate it?

Isn't the breakdown of consciousness (which is strategically speaking the reductionistic paradigm of Crick, Churchland and others) precisely the problem, since it breaks apart the very thing that must be experienced as a gestalt?

Consciousness isn't a thing to be described among other things, since it is the context, not the content, of what is experienced.

The Mobius strip is, by definition, an endless surface and any attempt to reorient it (by definition again) transforms it into that which it is not. Isn't consciousness akin to this very definitional paradox? My very attempt to correlate it or analogize it or reduce it or explain it simply upends a genuine understanding of what it is to experience it, since consciousness isn't a thing to be described among other things, since it is the context, not the content, of what is experienced. More precisely, consciousness cannot be exported as a piece of content, since it is a whole context in which such appearances arise. My attempts to reduce that holism to parts ipso facto means that whatever follows will be irretrievably "lost" in translation. If I understand Sam Harris correctly, I think this is what he is driving at when he writes,

"The problem, however, is that the distance between unconsciousness and consciousness must be traversed in a single stride, if traversed at all. Just as the appearance of something out of nothing cannot be explained by our saying that the first something was 'very small,' the birth of consciousness is rendered no less mysterious by saying that the simplest minds have only a glimmer of it."

Or, in more paradoxical (but illuminating) phrasing, Harris contends that we are merely deluding ourselves if we think we have solved the mystery of consciousness by our sleight of hand language shuffling. Harris writes,

"Likewise, the idea that consciousness is identical to (or emerged from) unconscious physical events is, I would argue, impossible to properly conceive—which is to say that we can think we are thinking it, but we are mistaken. We can say the right words, of course—"consciousness emerges from unconscious information processing." We can also say "Some squares are as round as circles" and "2 plus 2 equals 7." But are we really thinking these things all the way through? I don't think so. Consciousness—the sheer fact that this universe is illuminated by sentience—is precisely what unconsciousness is not. And I believe that no description of unconscious complexity will fully account for it. It seems to me that just as "something" and "nothing," however juxtaposed, can do no explanatory work, an analysis of purely physical processes will never yield a picture of consciousness. However, this is not to say that some other thesis about consciousness must be true. Consciousness may very well be the lawful product of unconscious information processing. But I don't know what that sentence means—and I don't think anyone else does either."

Tellingly, when Sam Harris attempted to convey such a sentiment to Daniel Dennett (the third—or is that the second?—atheistic horseman), the latter bemusingly responded that "If I can't imagine the falsehood of the above statement, I'm not trying hard enough. However, on a question as rudimentary as the ontology of consciousness, the debate often comes down to irreconcilable intuitions. At a certain point one has to admit that one cannot understand what one's opponents are talking about."

Perhaps Wittgenstein should make an entrance here and suggest that the real difficulty in studying consciousness is a language issue and that some things simply cannot be addressed by our symbolical logical systems. Was Wittgenstein providing us with a prescient warning shot on studying self-reflective awareness when he concluded his 1919 *Tractatus Logico-Philosophicus* with, "Whereof one cannot speak, thereof one must be silent."?

Or was Sir Arthur Eddington echoing a similar observation when he wrote, "All through the physical world runs that unknown content, which must surely be the study of our consciousness. Here is a hint of aspects deep within the world of physics, and yet unattainable by the methods of physics. And, moreover, we have found that where science has progressed the farthest, the mind has but regained from nature that which the mind has put into nature. We have found a strange footprint on the shores of the unknown. We have devised profound theories, one after another, to account for its origin. At last, we have succeeded in reconstructing the creature that made the foot-print. And Lo! It is our own."

Perhaps the study of consciousness has an inherent limitation, similar in import to Heisenberg's uncertainty principle in quantum mechanics or Godel's incompleteness theorem in mathematics. Perhaps we are like seasoned travelers on a Mobius strip in quest of the "other" side of the band who after long and arduous circular travels come to realize that no matter what route we take we will always only be touching the same surface. If this is so, then a specialized version of Niels Bohr's complementarity may be an instructive insight for us as we venture forth: "In any given situation, the use of certain classical concepts excludes the simultaneous meaningful application of other classical concepts."

In the study of consciousness it appears we may have to confront an epistemological complementarity where any objective study (via third person analysis) of qualia must by necessity lose in translation a fundamental feature of the very phenomenon under inspection. Conversely, any purely subjective endeavor to explore consciousness must by its very act forego any attempt to maximally objectify what is experienced, lest the experience itself be lost in attempting to exteriorize that which is de facto interior.

A broken down melody is, to quote one distinguished musician, no longer a melody. Similarly a broken down consciousness is no longer itself and therein lies the Sam Harris dilemma.

FOR FURTHER READING:

Sam Harris, "The Mystery of Consciousness", *http://www.samharris.org/blog/item/the-mystery-of-consciousness/*

MSAC Philosophy Group, "The Body That Surfs: *http://www.magcloud.com/browse/issue/287409*

Niels Bohr, "On Complementarity"

8 | *Maya: The Physics of Deception*

Every present moment is in truth a past moment, since what we think happens right now actually occurs nano seconds before. Even the experience of seeing this in front of your eyes is bounded by how fast the light can bounce off the screen and scatter back to your eyes which though exceedingly fast is lagged once the photons reach your optic nerve and then become transformed as chemical-electrical signals passing through a gauntlet of neurons and synaptic clefts to reach the visual cortex only to be recognized as an idea that was mistakenly believed to be instantaneous. Nothing arrives on time; rather everything arrives *in* time, even if we have been neurologically tricked into believing otherwise.

That the world is not as it appears is an ancient realization. Indian philosophy has captured this understanding in one simple, but nevertheless beguiling, word, *Maya*. There are several definitions of this Sanskrit term, which arguably first arose in the Vedic period of India ranging from illusion to magic. But perhaps the word's more literal etymology contains the most revealing explanation: "not that."

Maya in this sense means that which betrays its real origin and thus tricks us at each and apparently every turn into believing something about an event's causation that it not true. We see a sunset at a favorite beach and our experience is that it is happening right then, but we have discovered from astronomy that it actually takes eight minutes or so for the light from the sun to reach us. The same holds true with light from distant stars that travels thousands of light years to hit earth. We are not seeing the stars as they are but as they were hundreds and thousands of years ago. The world we behold is not so much in front of us in a persistent now as it is behind us in an escaping past. Thus we are continually, even if unconsciously, remembering our lives from transpiring experiences that blind us from their real origination.

It is as if the world is forever at a tilt but of which we remain unaware. I remember when I was a young teenager playing an old pinball game at the Fun Zone in Balboa which only cost a nickel but which continually gave replays. I thought I was both exceptionally skilled and lucky to play for over two hours on just one nickel. It was only later that I realized how mistaken I had been. The pinball machine was tilted in such a way that it provided constant replays to whoever was fortunate enough to play it.

Our brains are tilted in seeing the world at large a certain way, but in so doing it doesn't immediately inform us of this requisite fact. Even when I am supposedly making a conscious decision of whether to go right or left in my car, unconscious processes (for which I remain dutifully unaware) are determining my eventual turning of the wheel. Yes, I may believe that I am (and the pun here is intended) in the driver's seat, but a closer inspection of how we make choices at the neuronal levels illustrates the opposite. The neuroscientist John-Dylan Haynes conducted a study at the Max Planck Institute for Human Cognitive and Brains Sciences in Leipzig, Germany, which startlingly suggests that some decisions are unknowingly made on our behalf up to 10 seconds in advance. In this context, it is akin to the *Autopia* ride at Disneyland that gives children the illusion that they are steering the wheel and actually driving their respective cars, neglecting that they are on a track guided by a middle metal bar which automatically moves all vehicles along a preset design.

Deception is part and parcel of nature and is an intrinsic and necessary feature of human existence. Without it we wouldn't be able to survive, since survival of the fittest is predicated on one's ability to be stealth when necessary or to be able to invoke varying camouflages in times of need. As one evolutionary thinker put it, "Our brains were not designed to understand the universe as it is, but rather to eat it." And anything that can augment our eating habits (read survival skills) will be naturally selected, including our ability to deceive others. What is not so obvious, however, is that nature has also jerry rigged that the greatest deception of all will be within us. We cannot even look ourselves in the mirror without being subject to a fundamental illusion where that which is left is exchanged right.

Our five senses don't reveal the universe at large so much as provide us with a severely edited version of what our bodies necessitate to live long enough in order to pass on our genetic histories. For example, my hearing, smelling, and seeing is only within a certain frequency range and if something lies beyond that prescribed border then it remains unknown and non-existent to me. Thus the world I live in is an exclusive smattering of all that is possible, which is perhaps why John Lilly's famous witticism is more literally true than we might at first suspect, "In the province of the mind, what is believed to be true is true, or becomes true within certain limits to be learned by experience and experiment. These limits are further beliefs to be transcended. In the province of the mind there are no limits." The real question that arises, however, is not whether the mind is fully limited in some ontological sense (as Lilly questionably posits), but rather and more telling why there are limits in the first place.

Aldous Huxley, of course, famously argued that the mind was a filtering mechanism and that human consciousness simply couldn't function if it were allowed to receive all incoming data streams. If such a situation were to occur, perhaps our only viable response would be one of perpetual catatonia.

The ability to function for a set duration in any relative geometric space means that one must be largely edited by the physical rules governing such an enclosure. Hence, my psychology is the product of the intersection of physics and biology, even if I may be subjectively oblivious of how these laws actually work. The phantom limb sensation is a good illustration of just how mistaken we can be about what occurs within our nervous system. A majority of amputees report that they can still feel pain in their amputated hand, arm or leg, even when they know that such extensions have been long removed. The brain is a modeling system and even if an arm has been amputated the image of that arm (and its attendant connections to pleasure or pain) still resides within the Rolodex of remembered sensations. We can even see things that are not there. Perceptual "filling-in" is how our spectra of vision compensates for missing information due to our physiological blind spots. Thus a wobbly image can be arrested and stabilized on our retina which can lead to a filling up of the surrounding background by images that are literally not there.

Ronald Siegel, a distinguished researcher on the effects of psychotropic drugs on altered states of consciousness at UCLA, described in a remarkable study how a series of participants who were given a hallucinogenic dose of marijuana and other drugs all ended up reporting seeing one recurring and disturbing image of multiple eye balls staring back at them. At first Professor Siegel couldn't explain why each of his "psychonauts" had the same experience, but on closer inspection he realized that in their initial training session each of the drug takers had watched a slide show which accidentally contained a psychedelic portrait of eyes.

As Siegel explains it, "The near-toxic dose of mescaline I had ingested by drinking peyote all night kept the Demon alive for many seconds, long enough "to see." The eyes looked like pictures that had been cut out of magazines and pasted together in a collage conceived by a deranged artist. In the lower right-hand corner saw the letters ES followed by a series of numbers. The Demon faded away before I could read them. But I had seen enough. The letters and numbers were part of a code I put on the borders of the slides used in the psychonaut training course. The subjects never saw the code numbers, but I used them for identifying the contents of each slide. These slides were projected on a piece of black cardboard (the black curtain) tacked to the laboratory wall. The training slides were all black and-white drawings of simple geometric forms such as tunnels or lattices. The ES series was very different. It consisted of dramatically colored 'psychedelic' scenes created by artists for light shows, Hollywood films, and other commercial productions. I had obtained a collection of these slides from Edmund Scientific, a mail-order supply house in New Jersey. But I had not used the ES series in training. I was saving them to show the psychonauts after the experiments were completed so that they might be able to select images that were similar to their own hallucinations. Somehow, one of the ES slides must have slipped into the batch of training slides and imbedded itself in the psychonauts' memories. I was certain that when I returned to the lab I would find it."

And find it, he did. What Professor Siegel realized was that, "The Demon was nothing more than the surprise of a disturbing image spontaneously retrieved from memory. Rather than feeling disappointed that a 'real' Demon did not exist, I was surprised and humbled to discover that internal images can be powerful enough to be mistaken for external ones. Disturbing images have a way of burrowing their way into our memories, even after a single exposure."

Freud had long believed that one of his great, if not greatest, discoveries in psychology was how transference amongst some of his patients operated. Within about six months or less, they would project (albeit unconsciously) all sorts of fantastical ideas about their doctor but without a hint of recognition that it was their own doing. In other words, they took to be objective that which was in truth wholly subjective.

This conflation of one's internal brain state for an objective reality is an elemental part of what it means to be human. While most of the time such a correlation can be of great advantage (particularly when such overlays are mostly accurate and predictive), at other times it leads to massively delusional states of awareness. A recurring pattern of such delusions are our own dreams which we habitually take to be deeply real and rich with episodic narratives, except when we wake up and soon realize their wholly imaginative nature. But because consciousness is a virtual simulator there are times that our dreaming brain can overlap with our waking state and radically confuse us about what is internal and what is external. In certain brain states it is nearly impossible to even recognize our own projections as projections.

Near-Death Experiences are a good case in point here. The luminous out of body experiences with their accompanying visions can be so intense that it is nearly impossible to question the veridicality of any religious vision that may arise within one's purview. Yet apparitions of Jesus only appear to Christians, and Buddha only to Buddhists, and Guru Nanak only to Sikhs, and Krishna only to Hindus, which should give anyone pause about the objectivity of all such manifestations. Moreover, the one undeniable factoid about NDE's is that the person didn't die but rather lived long enough to retell his or her tale. From an evolutionary perspective, this seems to indicate that NDE's are not about a purported afterlife but rather about the brain's amazing ability to create a reason or purpose to continue living, drawing as it does from the person's own unique biographical circumstances. Simply put, NDE's are projections of a person's ultimate concern and those concerns (for better or worse) invariably motivate one to live another day. But such a mechanism isn't very effective if one doubts its numinous origins while undergoing the transformative encounter. In other words, the brain tricks us into believing its own machinations as something that is not sui generis. All this trickery does serve one underlying purpose: keeping our organism intact long enough to recapitulate itself.

The very colors we perceive are not a product of some Kantian insight into the electromagnetic spectrum itself, but rather the very opposite of what our common sense intuits. A blue sky isn't blue because the air is permeated with blueness, but because of the different wavelengths of light. Shorter wave lengths of light tend to get absorbed by varying gas molecules and as such then radiate in different directions around the sky and thus that very light reaches our eyes and we then see a blueness everywhere. But there is no such thing as "blueness" as a thing in itself, try as we might to capture it within the palm of our hands.

Science, contrary to some popular definitions of it, is not common sense realized. Rather, it is oftentimes a counter-intuitive procedure that necessitates a new way of looking at the world. This may explain why even in the 21st century so many people resist the implications of evolutionary biology and favor more story-laden narratives (replete with a meaningful plot) such as intelligent design.

Nothing is as it seems. Ancient rishis argued that the world was an illusion and that what we take to be real isn't. Plato's allegory of the cave and the Gnostics idea of demiurges also touch upon this perennial insight. The ultimate goal for both was to wake up from this continual dream into a greater reality which will reveal a higher truth. Of course, the very idea of a higher truth may itself also be a progressive form of deception in higher mammals to help evolve better strategies to keep certain genetic histories alive and kicking.

In this way the Hindu idea of Maya is one of an all-persuasive goddess, most popularly known as Lakshmi, who has the power either to ensnare a soul in samsara or to liberate her from the cycle of reincarnation and karma. It is a deep irony that millions worship the very incarnation of deception. Of course, it can be argued that everyone is in some ways a devotee of Maya, since we all live within her bewitching web, where confusing cause and effect and image and object are our natural habit. And, yet maybe understanding the very root of Maya is precisely how and why science has progressed where other paths have failed. Why? Because science, unlike traditional religions of the past, is a consistent method of doubting what we think we know and not relying on what we may wish to be the case.

Science, in other words, is a path of systemic disbelief, which like the very literal meaning of Maya, looks askance at disparate phenomena (and its attendant explanations which invoke affirmations of "I believe, I believe") and proclaims something inherently more radical, neti, neti: "not this, not that." And by such a procedure patiently unravels the hidden knot that for millennia has remained unraveled. *I became free of Maya when I realized that I could never escape from her.*

103

"Although we experience the illusion of receiving high-resolution images from our eyes, what the optic nerve actually sends to the brain is just a series of outlines and clues about points of interest in our visual field. We then essentially hallucinate the world from cortical memories that interpret a series of movies with very low data rates that arrive in parallel channels."
—Ray Kurzweil, *How to Create a Mind*

"If we accept that consciousness can be simulated, at least in principle, it is then only a small step to imagining that something like a conscious human being could be simulated."
—Paul Davies, *The Cosmic Jackpot*

"Unless we are now living in a simulation, our descendants will almost certainly never run an ancestor-simulation."
—Nick Bostrom, *Philosophical Quarterly*

"Even though we think we see the world so fully, what we are receiving is really just hints, edges in space and time."
—Frank S. Werblin

The idea that the world is an illusion that betrays its real origin has a long tradition and can be found in the writings of Hindu rishis, early Greek philosophers, and Christian gnostics. What is perhaps surprising is to find such a rich literature on the subject in neuroscience and quantum physics.

The latest, and perhaps most provocative, idea to gain some currency in varying scientific disciplines is the hypothesis that the universe is the result of a computational simulation and, as such, is an incredibly rich and detailed illusion which has ultimately tricked us into believing otherwise.

Nick Bostrom's now famous 2003 essay, "Are You Living in a Computer Simulation?" has articulated the concept very simply with three propositions, wherein he argues that at least one of which will turn out to be true.

"(1) The human species is very likely to go extinct before reaching a "posthuman" stage; (2) any posthuman civilization is extremely unlikely to run a significant number of simulations of their evolutionary history (or variations thereof); (3) we are almost certainly living in a computer simulation. It follows that the belief that there is a significant chance that we will one day become posthumans who run ancestor-simulations is false, unless we are currently living in a simulation."

While Bostrom's argument has been met with resistance from a number of quarters, it is intriguing to see how it dovetails with Hugh Everett's "Many-Worlds" interpretation of quantum mechanics which is gaining some traction among theoretical physicists, particularly as it has been coupled with M-theory and the far reaching notion of a multiverse.

Neuroscience has more or less established that the brain is a simulator par excellence and that what we see, hear, touch, and smell are the results of how our central nervous system processes both external and internal stimuli and then reconstructs a virtual environment in which we react accordingly. Simply put, the reality we experience is part and parcel a simulation and may or may not correlate (at differing times and differing places) with what we believe exists externally from ourselves. We already intimately know that the brain is a virtual simulator because of dreaming where everything is hallucinated by us, even without us knowing how and why we are doing it. The waking state differs from dreaming because it receives external data streams from the nine orifices of our body, which allows for new material from which our brain can draw new maps about how to respond to any given situation.

But in both cases—dreaming or waking—we are living in a simulation created by a neural network that has billions of on/off nodes tied in with trillions of synaptic clefts, all of which creates worlds upon worlds within our own skulls, even as we employ such models to interact with others of similar or dissimilar modeling dispositions.

Arguably, consciousness is a virtual simulator, apparently evolved over eons of time to enable mammals with higher brain functions to "in source" varying options of how to respond to a disparate array of problems before "out sourcing" them (hence the Darwinian advantage).

The proviso in such simulations, however, is how real they feel when occurring. To the degree that such paralleling seems certain and solid we react accordingly. However, if such replicas appear imaginary we tend not to engage them seriously and interact with them in multifarious ways. The dividing line between what we take to be reality and fantasy is a contingent and moving one.

This is perhaps best illustrated in the world's religions where one's person's version of transcendental truth is oftentimes viewed by others as a manufactured delusion or worse as a product of a nefarious demiurge.

Faqir Chand, the famed unknowing sage of Hoshiarpur, has revealed how religious visions though appearing miraculously real are in essence illusory projections of one's own faith and belief. It was in Basra Bagdad (Iraq) during World War One when Faqir realized the pivotal secret in understanding transmundane phenomena and how easy it is for our consciousness to be tricked by neural mock-ups. In his autobiography, *The Unknowing Sage*, Faqir relates how in the middle of a battle at Hamidia the form of his guru Shiv Brat Lal manifested to him and said, "Faqir, worry not, the enemy has not come to attack but to take away their dead. Let them do that. Don't waste your ammunition." Faqir then sent for the Subedar Major and narrated the appearance and direction of his guru. He followed the same strategy and all were saved. When Faqir reached Bagdad after the fighting, however, many of Shiv Brat Lal disciples began to worship him instead.

Faqir recollected: "It was all unexpected and strange for me. I enquired of them, 'Our Guru Maharaj is at Lahore. I am not your Guru. Why do you worship me?' They replied, 'On the battle field, we were in danger. Death lurked in hand. You appeared before us and gave us correct directions. We were spared.' I was wonder struck by this explanation. I had no knowledge of it at all. I, myself, being in trouble at that time,

107

had not even remembered them. A mystery shrouded the whole thing, 'who appeared inside them?'"

When Faqir discovered that his own guru (Shiv Brat Lal) was unaware of his manifestations, he concluded that the answer to the perplexing problem of religious visions must rest in the nature of consciousness itself.

Faqir elaborated: "People say that my Form manifests to them and helps them in solving their worldly as well as mental problems, but I do not go anywhere, nor do I know anything about such miraculous instances. O' Man, your real helper, is your own Self and your own Faith, but you are badly mistaken and believe that somebody from without comes to help you. No Hazrat Mohammed, no Lord Rama, Lord Krishna, or any other Goddess or God comes from without. This entire game is that of your impressions and suggestions which are ingrained upon your mind through your eyes and ears and of your Faith and Belief."

Thus, following Faqir's lucid argument, the modus operandi for religious visions is not due to outside or disconnected forces (although exterior stimuli can act as a catalyst for it), but to the internal process of concentration. A force that for approximately sixteen hours a day enables one to see the everyday, common sense, lawful world, and for another several hours at night can allow one to fly to the moon, converse with unknown people, and create incredible panoramas. Consequently, the appearance and duration of such visions is intimately related to attention and focus.

Dreaming serves as the classic and perhaps most misunderstood example. The nature of one's attention is related directly to the perception one experiences. If our perspective alters so does what we perceive. As ancient Upanishadic speculation and current studies in consciousness have shown, we do not see the world as it "is." Rather, owing to our neurological structures, we see the universe — incoming stimuli — relatively; appearances flowing in and out depending on our own biologically defined anatomies. This "predicament" has meaning, content, and purpose within the framework of our own lived-through experiences. However, it is naive to say that our interpretation of life from science, philosophy, or religion absolutely explains the world as it really is. Instead, what we have are metaphorical models of explanation, which work respectively within the brackets of our own curtailed existence. The unseen thread, the larger gestalt, however, will go by undetected. With sharply contoured (mathematical, if you choose) operating mechanisms, we find ourselves living in a universe understood not by pure perception but by alternating analogs.

What these metaphors are (or, more precisely, which limited stream of reality we behold) depends on what I call the *Chandian Effect — the experience of certainty,* named after the late Faqir Chand who was the first person in the Sant Mat tradition to bring this issue to light. It is from this bedrock quality that we distinguish, acknowledge, and discriminate so-called reality from appearance or illusion. What we call the "actual" world is dependent solely upon the vibration and consistency in the persuasiveness of certainty.

Although we can see, hear, smell, and touch our reality, what determines our conviction that this world is real is not so much based upon objective datum as it is on our subjective "feeling" (even if chemically moderated) of certainty.

The experience of certainty is a propelling force behind how we make up our days, fashion our plans, articulate our hopes. If there occurs a break in the *Chandian Effect* (sleep too little or drink too much, for examples) our normal waking state collapses into a passing phantasm. Like our nocturnal dreams, it gets stored away and temporarily forgotten. The experience of certainty is so overwhelming that when it radiates forth the question of illusion seldom arises. Just as the chair is quite solid when we strike it with our hand, so too does the world appear concrete and vivid when the *Chandian Effect* pervades.

But dreams can on occasion seem as certain as anything in the waking state and sometimes in lucid dreams or near-death events even more luminous than anything we have experienced before. If both dreaming and the waking state are the result of our brain simulating internal and external stimuli, then what makes us certain of either reality is, ironically, also a simulation and the distinction between the two is not as clear as we might suspect.

Our state of reality is determined by the movement of consciousness into various expressions of the *Chandian Effect*. Each level of awareness is controlled and empowered by the degree of certainty we experience, which is determined by the intensity and duration of its minimum threshold, which is precisely the point at which one state (e.g., the waking state) blurs or fuses with another state (e.g., dreaming). The single most obvious example is whenever we try to stay awake for more than two or three days at a time and being deprived of sleep our dream world intrudes upon our otherwise "sane" reality. We are predisposed to call the waking state "real" because it is longer (and hence, by extension, more vivid) than the dreaming stage. Yet, we generally say this only when we are awake but never while we are dreaming. The reason behind this is simple. At each level where attention is established (or, neurologically speaking, whenever key neural triggers are operative), a certainty boundary is in effect, which, owing to the given center of awareness, varies in strength, time, and permanence. Hence, even the waking state, although seemingly vivid, only lasts about eighteen hours normally until the *Chandian Effect* (or neural parameters) structured upon this level runs down below the minimum threshold and our consciousness shifts to another region. So it is with the dream stage. At the moment of sleep (itself nothing but the transition of attention) we find ourselves occupied in a world that just hours before we thought was nothing but an incredible illusion — because it was dimmed by the intensity of the certainty force inherent in the waking state — but with which we now deal quite seriously: running away in terror from death or luring attractive mates for orgasmic satisfaction. From this native pattern of awareness we can see that our lives are simply natural progressions of consciousness from various boundaries within the *Chandian Effect*.

All of this leads us to what appears to be an inescapable conclusion: we apparently cannot adjudicate the ultimate reality of any stage in consciousness on the basis of how certain or real it may appear since any simulation can magically seduce us into believing its superior ontology. We are, it seems, in an intractable position within any stage of awareness to gauge its ontological reality if such a reality is indeed the product of a series of simulating intersections. At best we can simply argue about the relative hierarchical features of differing simulations, with the added caveat that such rumination is also circumscribed within a parallelism of its own.

This is perhaps why the idea of our universe itself being a computational simulation is not as far fetched as we might imagine, given the replicating features of mammals who possess a neocortex. Given enough computational power and given the ability to embed matter with unimaginable complexity, it isn't implausible to envision a world that for all intents and purposes looks just as real (if not even more vivid) than the one we currently occupy. Indeed, as Vernon B. Mountcastle points out in *The View from Within*,

"Each of us lives within the universe—the prison—of his own brain. Projecting from it are millions of fragile sensory nerve fibers, in groups uniquely adapted to sample the energetic states of the world around us: heat, light, force, and chemical composition. That is all we ever know of it directly; all else is logical inference."

We have now advanced far enough in our film technology to be able via CGI to recreate on screen the most intricate features of various animals and landscapes to such a degree that audiences worldwide can be so convinced that a fake tiger is so real that even when the illusion is explained they can still vehemently argue otherwise, as what has recently happened with the movie version of the *Life of Pi*.

Given this unusual state of affairs, we can also now be deceived into mistaking a CGI effect when it was in actuality a genuine phenomenon. A good example of this reverse Turing effect can be seen by the comments made to a YouTube video entitled Tsunami surfing where Mike Parsons (a real surfer) rides a huge (and real) wave in Maui, but which many commentators believed was created artificially. In the world of film, computer simulations have already passed the

111

Turing test where that which is artificial is literally taken to be real. One can only imagine what cinema will look like in three decades hence.

Is this universe of ours an app similar to a game like Spore, where creatures confronting competing species evolve over time, which can be downloaded for free on an iPhone or an iPad? Before we dismiss such conjectures as silly science fiction speculations, the more viable question we should ask first is how difficult will it be in the future to computationally recreate the world we now experience, given the exponential growth of computing power and the law of accelerating returns? In addition, how difficult will it be to trick us into believing in the certainty of any given state created by artificial intelligence?

The answers may surprise us. It is probably much easier than we at first would suspect, especially in light of how malleable human consciousness can be.

Faqir Chand once related how when he was a young Brahmin growing up he had daily visions of Lord Krishna. One day as he was walking on his way to town Lord Krishna appeared to Faqir and told him to eat some cow dung which was on the street. Faqir did as Krishna requested, but later on when the vision had disappeared and when Faqir was in a more rational-logical state of mind, he reflected that he had never heard or read of any god or goddess asking a devotee to do such a thing. This caused Faqir to doubt the reality of the apparition and he stopped having visions of Krishna.

This incident and others similar to it suggest that the modus operandi for any reality to lose its *Chandian Effect* one must severely doubt its fundamental certainty, whether by chemical augmentation or by radical and systematic skepticism. Of course, some states seem more recalcitrant to questioning than others and thus give the impression of being "more" real than others. But given that every state we experience is a simulation, the hierarchical nature of more real vs less real seems bounded by the necessity of survival within any given region of awareness.

Indian philosophy, particularly some schools of Advaita Vedanta, have long argued that the universe really is a play or a game which is commonly called lila or leela (literally, "pastime or sport"), not dissimilar to applications one can download on Apple or Android devices. Darwinian natural selection would appear to be one of the key operating systems of this universal sport.

Stephen Wolfram, the architect of Mathematica and founder of Wolfram-Alpha (the computational knowledge engine), believes that a new kind of science has been born with our understanding of how incredibly complex systems can be algorithmically reconstructed by computational reducibility. If this is true, then perhaps Nick Bostrom's propositional question isn't merely a fanciful and sophisticated reworking of the Matrix's underlying premise. Perhaps Faqir Chand's skepticism of religious visions (and there eventual dissolution because of it) is an indicative pathway by which to question the certainty of any state of awareness that may arise. Science has made tremendous progress by doubting erstwhile commonsense explanations, even to the point of questioning the very reality of all that we take to be solid and permanent.

Underlying science and almost all human endeavors is the supposition that the world we experience (even if by neural trickery) is the supreme reality by which all other realities are (and should be) measured. But if Bostrom's third supposition is true (that we are already living in a computational simulation) then it might be illuminating to question the very foundational basis of our waking state certainty. Clearly, accepting the Hindu idea of reality as an illusory sport played out by the gods allowed rishis in the past to develop deep philosophical insights into the nature of the mind long before the advent of neuroscience. Similarly, imagining that the universe is an app or a matrix like mock-up designed to deceive its participants from knowing its real causation can serve as a powerful and enlightening awakening, even if such conjectures turn out not to be wholly true or accurate.

Why? Because any idea that can jar us from complacency and force us to think anew about reality is helpful to a consciousness that evolved to virtually simulate reality and play out competing scenarios.

As Frank S. Werblin so wisely points out, "Even though we think we see the world so fully, what we are receiving is really just hints, edges in space and time." In a universe with an almost infinite set of data streams, we as Homo sapiens should be exceedingly cautious in believing (as we invariably do) that 10 or 12 channels in this spectrum are sufficient to inform us about the sum total of reality. We are sparse coders indeed who have been neurologically duped by evolution and natural selection into believing that our consciousness is a transparency when on closer inspection it is more akin to an elaborate labyrinth with no decipherable way of knowing where it begins and where it ends. Or, as Jorge Borges once put it, "You have wakened not out of sleep, but into a prior dream, and that dream lies within another, and so on, to infinity, which is the number of grains of sand. The path that you are to take is endless . . . ."

It is easy to have visions when one is meditating, provided one hasn't slept for a couple of days. Hypnagogia, which is usually defined as the "transitional state to and from sleep," can induce all sorts of fantastic apparitions and strange admixtures of light and sound. However, how one ultimately interprets these passing phantasms seems to be directly correlated with theoretic prefiguring.

John Lilly, who was famous for his pioneering studies on how consciousness behaves when deprived of incoming stimuli, came to realize that preset modeling played a transformative role in how one ultimately viewed inner experiences, whether induced by spiritual practices or hallucinogenics. His findings indicated that there is was an almost intractable problem confronting the scientific study of the mind because, "your theories or explanations will determine which experiences you will or will not have, no matter what experiments you perform. It is difficult to test a theory in this realm if other beliefs limit the range of available experiences."

For instance, on occasion when I meditate at night and I haven't had sufficient sleep I have very lucid visions of all sorts of phenomena. They appear just as real as anything I witness in the waking state, but these experiences are transpiring with my eyes completely closed. Once I saw a wonderful cascading shower of rain, filled with different colors, and it literally felt as if water drops were bathing my face. Phenomenologically speaking, I would be hard pressed to differentiate this hypnagogic state from what my normal waking awareness.

Yet, I am quite familiar with visual hallucinations and thus I don't take such liquid displays too seriously and I certainly don't give them any special meaning. However, and here is where Lilly's understanding of theoretic prefiguring looms large, if I followed or believed in a different paradigm which took such lucidities as signposts of a higher spirituality (such as with Eckankar or MSIA which places a high value on lucid dreaming) I would most likely interpret such phantasms in a more positive and significant light.

John Lilly provides us with a telling example, as recounted in his autobiography, "In profound isolation, one may have the experience that other people are present, or that one is receiving communications from sources outside the tank. What one makes of these experiences depends on one's beliefs. A person who believes in telepathy is likely to conclude that the messages are actually coming from someone or something at a distance from the tank. Within such a belief system, the experience will be perceived as real. On the other hand, someone who does not believe in telepathy, having the same basic experience, will perceive the experience as unreal, perhaps calling it an 'hallucination.'"

Interestingly, when John Lilly had mystical experiences of what he perceived as other worldly beings in his sensory deprivation tank (after taking mind-altering drugs, such as Ketamine), he argued strongly for their ontological objectivity. Whereas, Richard Feynman, the Nobel prize winner in physics for his work on quantum electrodynamics, who had similar experiences as Lilly (they were friends) labeled his out of body excursions as "hallucinations." Feynman's categorical dismissal miffed Lilly who then criticized him in a personal letter with the terse rebuttal, "you stopped being a scientist the instant you said that word, hallucination."

Feynman countered Lilly's assertion by pointing out that whatever out-of-body experiences he was having (and Feynman recalls becoming very good at) didn't correlate to the outside world, even when undergoing the dissociation he thought they did. It was for this reason that he tried to convince Lilly that "the imagination that things are real does not represent true reality. If you see golden globes, or something, several times, and they talk to you during your hallucination and tell you they are another intelligence, it doesn't mean they're another intelligence; it just means that you have had this particular hallucination.... I believe there's nothing in hallucinations that has anything to do with anything external to the internal psychological state of the person who's got the hallucination."

Feynman's viewpoint, of course, is also underlined in the *Bardo Thodol* (or more famously known in the West as the *Tibetan Book of the Dead*), which also describes the illusory nature of such inner encounters. Evans-Wentz summarizes the Buddhist viewpoint thusly, "'That all phenomena are transitory, are illusionary, are unreal, and non-existent save in the sangsaric mind perceiving them. . . That in reality there are no such beings anywhere as gods, or demons, or spirits, or sentient creatures—all alike being phenomena dependent upon a cause. That this cause is a yearning or a thirsting after sensation, after the unstable sangsaric existence."

John Lilly countered Feynman by arguing, "that the word hallucination is a trash-bin concept for a whole range of experiences that people wish to discount because they are unconventional or difficult to describe. The term is an unscientific generalization that confuses a multitude of significant processes and specific experiences involving internal reality."

This apparently indissoluble impasse (mysticism vs. neurology? or superluminal vs. mundane?) has implications that are more far-reaching and paradoxical than one might at first suspect. A pregnant illustration of this is how a limited purview can actually tripwire an erstwhile rationalist into believing something contrary to their usual commonsense.

As Lilly explains, "A behaviorist, usually the most skeptical and hard-boiled sort of psychologist, might enter the [deprivation] tank with the belief that nothing can happen in the brain without some external stimulus as the cause. Should an experience occur which appears to come from a source outside his own head, but with no such possible source nearby, such a person might then be forced to adopt a belief in telepathy in order to explain it. Within the belief system of the behaviorist, this becomes the only acceptable explanation for such phenomena in a framework that does not allow for any inner experience or even hallucinations. Paradoxically, then, a person who does not even believe in the psyche may end up believing in 'psychic communication.'"

What is so unusual about this theoretic prefiguring is that, if it not closely guarded and checked, it can have devastating personal and political consequences, such that a relatively normal and sane person could be diagnosed as psychotic not because of their experiences per se, but by not proffering the politically correct map by which to adjudicate it.

As Lilly insightfully foretells, "Any normal, healthy person could have unusual experiences that would seem real if they believed in them [but] if he did not believe in them, the experiences would seem unreal." In either case the experience is the same, but the interpretative model is different in each. If, however, "such a person believes in the reality of the experience and communicates this belief to a psychiatrist, the psychiatrist would conclude that the individual was hallucinating and might be psychotic."

However, on the other end of the spectrum (as Lilly point outs) if the patient describes an unusual experience and professes not to believe in its reality, then this same "psychiatrist might be less inclined to assume mental pathology."

This then leads us to a most unusual "neural paradox" wherein one is judged not by the experience itself but rather by one's "beliefs or opinions about his experience." Yet in most cases these set of beliefs (or more properly worldviews) were already in place before one had any inner experiences whatsoever. Lilly soon realized after his fantastical encounters of other beings and other universes (frighteningly retold his classic book, *The Center of the Cyclone*) that how one recounted the experience (as a believer in their reality or as merely an unreal hallucination) made all the difference in how the outside listener judged his relative sanity.

More importantly, however, was how one personally regarded such experiences, since some inner journeys were so utterly horrific that those who saw them as merely hallucinatory night terrors could liquidate their gnawing fears by recognizing their dream-like and unreal nature. But others who firmly believed that such excursions could be caught in a hellish nightmare from which it appeared impossible to escape.

118

As Lilly astutely summarized, "[Such] explanations and theories are really beliefs about the universe and the mind. A particular belief may or may not be true, may or may not cause one to act in a certain way, but that belief will unquestionably set limits on what one can experience."

How then can a science of consciousness proceed if at the very outset our neural prefiguring already contours our eventual adventures of the mind? John Lilly's answer to that query has now become his most cited philosophical witticism:

"In the province of the mind, what is believed to be true is true or becomes true, within certain limits to be found experientially and experimentally. These limits are further beliefs to be transcended. In the province of the mind, there are no limits."

In other words, if consciousness is a virtual simulator then it can potentially simulate anything given the necessary information. The glitch here, however, is that a simulating brain is invariably bounded by what it believes to be possible, particularly in a society which by its very structure tends to limit what can be plausibly accepted at any one particular point in time and space. Breaking through such psychological and cultural boundary lines is, of course, a daunting task since such taboo breaking makes one a potential outcast. It is also dangerous since there is a fine line between acting as if something "may be real" versus acting as if something "is indeed" real.

Richard Feynman cautioned his would-be scientists at Cal Tech on their graduation day that "The first principle is that you must not fool yourself--and you are the easiest person to fool." While this is certainly helpful advice, the problem is that our brains were designed to trick us from the very start since every experience we have of the world both within and without isn't as Kant pointed out centuries prior "the thing in itself" but rather the end-result of filtering process which invariably colors whatever we see and hear and touch around us.

The recognition that the mind is a simulator par excellence doesn't liberate us from its unceasing simulations, since even that recognition is part and parcel a simulation as well. We are living in a neural paradox which may be likened to an endless hall of mirrors where what we recognize as real and substantive may on closer inspection be merely an image of an image of an image, ad infinitum.

Arguably the key question that begs to be answered in our quest to understand consciousness is whether our self-reflective awareness can be algorithmically understood in terms of information theory (and thus is potentially substrate neutral) or, is, as Sir Roger Penrose has long argued, the result of quantum biological processes which cannot in principle be computationally reduced.

In the early 1960s Dean E. Wooldridge, a Research Associate at California Institute of Technology and a Director of TRW, laid out in his now prophetic book, *Mechanical Man: The Physical Basis of Intelligent Life*, strongly posited that "if the properties of consciousness can indeed be shown to be precisely determined in rigid cause-and-effect fashion by the physical state of the associated material, then conscious phenomena clearly belong to the subject matter of basic science. The unusual properties of consciousness, which make it seem so different from quantities that we think we understand better, do not disqualify it for inclusion. Indeed, if concepts had in the past been excluded from physics when they seemed too bizarre or hard to comprehend, there would certainly be no relativity or quantum mechanics today. . . . [As such we are] finally ready to make the same transition from metaphysics to physics that was set in motion for the other functions of the body in the early 1600s." (Pages 161-162).

It was not so long ago that many people, including some very eminent thinkers (such as Henri Bergson), believed that the secrets of genetics would never be revealed by biochemistry because there was something inherently non-reducible in life's coding system, something akin to supernatural vitalism. But this turned out to be spectacularly wrong when Francis Crick and James Watson discovered the double helix structure to DNA and how four basic building blocks, adenine, cytosine, thymine and guanine comprise the fundamental language in life's evolution. It is the reconfiguration of these letters (as base pairs A T and C G) which make up our genotypes and which in turn determine the elementary differences between a dolphin, a human, and a grain of rice.

What may have seemed unimaginable in the early 20th century (unraveling the billions of line of code of a human genome) is now viewed as relatively commonplace. Is it conceivable that the mystery of consciousness may also have an informational solution similar to the genomic revolution?

Christof Koch and Giulio Tononi believe that integrated information theory is the key to finally understanding how consciousness arises from complex chunks of matter. In this view, even though self-awareness is constructed from immensely varied, but connected, data streams, what we experience in our consciousness is integrated and thus differentiated parts come to us in a qualia gestalt.

PHI: A Voyage from the Brain to the Soul

Hypothetically, this means that the amount of information that any system can compute at any given time is correlated to a conscious state. As such, this indicates that even though reductionism can indeed unpack the varying incoming data waves, the subjective experience of consciousness is experienced as a whole and thus cannot be properly understood unless that totality is taken as a given. Another way of understanding this is that qualia may consist of any number of sub-routines that contribute to its subjective character, but it is when those pathways conjoin which leads to an integrated sense of consciousness.

Neuroscientist Giulio Tononi's *PHI: A Voyage from the Brain to the Soul* summarizes it this way, "Integrated information measures how much can be distinguished by the whole above and beyond its parts, and F is its symbol. A complex is where F reaches its maximum, and therein lives one consciousness— a single entity of experience."

INFORMATION SYSTEMS

However, Koch's and Tononi's integrated theory shouldn't be conflated with Ken Wilber's integral theory, since the former doesn't invoke a Spirit or Consciousness first principle, but rather focuses on how material complexity (the totality of specific informational patterns) is coincident with self-reflective awareness. As Koch illuminates,

"It's not that any physical system has consciousness. A black hole, a heap of sand, a bunch of isolated neurons in a dish, they're not integrated. They have no consciousness. But complex systems do. And how much consciousness they have depends on how many connections they have and how they're wired up."

Though this may at first glance appear to advocate a top down approach, it is more properly adjudicated as an informational scaffolding project, where very close attention is given to neural complexity and the vast material interconnections necessary that gives rise to varying levels of consciousness.

Thus it is the level of integration (not necessarily the compositional strata of that integration) that gives rise to awareness. This then suggests that consciousness is substrate neutral, which means that all sorts of non-organic material could potentially reflect consciousness, including the Internet. As Koch controversially admits,

"But according to my version of panpsychism, it feels like something to be the Internet — and if the Internet were down, it wouldn't feel like anything anymore. And that is, in principle, not different from the way I feel when I'm in a deep, dreamless sleep."

John Searle, who has been a professor of philosophy at U.C. Berkeley for more than five decades and is famous for his contrarian views on how consciousness should be studied, finds Koch's panpsychism to be unpalatable. In a scathing critique in the *New York Review of Books*, Searle writes, "But the deepest objection is that the theory is unmotivated. Suppose they could give a definition of integrated and differentiated information that was not observer-relative, that would enable us to tell, from the brute physics of a system, whether it had such information and what information exactly it had. Why should such systems thereby have qualitative, unified subjectivity? In addition to bearing information as so defined, why should there be something it feels like to be a photodiode, a photon, a neutron, a smart phone, embedded processor, personal computer, "the air we breathe, the soil we tread on," or any of their other wonderful examples? As it stands the theory does not seem to be a serious scientific proposal."

Both Koch and Tononi, however, argue that their theory is open to scientific refutation and thus qualifies as a serious scientific endeavor, contrary to Searle's doubts. In a recent issue of *Wired Magazine*, Koch was asked, "Is your version of panpsychism truly scientific rather than metaphysical? How can it be tested?" To which Koch responded, "In principle, in all sorts of ways. One implication is that you can build two systems, each with the same input and output — but one, because of its internal structure, has integrated information. One system would be conscious, and the other not. It's not the input-output behavior that makes a system conscious, but rather the internal wiring. The theory also says you can have simple systems that are conscious, and complex systems that are not. The cerebellum should not give rise to consciousness because of the simplicity of its connections. Theoretically you could compute that, and see if that's the case, though we can't do that right now. There are millions of details we still don't know. Human brain imaging is too crude. It doesn't get you to the cellular level. The more relevant question, to me as a scientist, is how can I disprove the theory today. That's more difficult. Tononi's group has built a device to perturb the brain and assess the extent to which severely brain-injured patients — think of Terri Schiavo — are truly unconscious, or whether they do feel pain and distress but are unable to communicate to their loved ones. And it may be possible that some other theories of consciousness would fit these facts."

Interestingly, and perhaps more revealing than one might at first suspect, the integrated informational approach has parallels, at least in part, with genomics and how complex

DNA strands when properly sequenced give rises to organic life. As J. Craig Venter brilliantly illustrates in his latest book, *Life at the Speed of Light: From the Double Helix to the Dawn of Digital Life*, "Life is an information system" and only when each part of that code is properly aligned can an organic cell properly function. Even the tiniest of errors (where just one nucleotide is misplaced) can have catastrophic consequences.

As Venter explained when trying to synthesize the M. mycoides Genome, "Having established which segment contained an error or errors that did not support life, we sequenced the DNA once again, this time using the highly accurate Sanger sequencing method, and found that there was a single base-pair deletion. If this sounds as trivial as writing 'mistke' instead of "mistake," equating nucleotides to individual letters is slightly misleading, in the sense that DNA code is read three nucleotides at a time, so that each three-base combination, or codon, corresponds to a single amino acid in a protein. This means that a single base deletion effectively shifts the rest of a genetic sentence that follows, and hence the sequence of amino acids that the sentence codes for. This is a called a 'frameshift mutation'; in this case, the frameshift occurred in the essential gene dnaA, which promotes the unwinding of DNA at the replication origin so that replication can begin, allowing a new genome to be made. That single base deletion prevented cell division and thus made life impossible."

GENOMICS

Is consciousness as an informational coding system (defined by its integrated sets of computational interactions) similar to genomics where the real emphasis must be on how individual parts (be it neurons or nucleotides) combine and reconfigure to give birth to integrated complexes (be it a living cell or a conscious entity)?

Interestingly, Venter and his scientific team have discovered by their painstaking and labor intensive efforts that life itself may also be substrate neutral, since they have successfully created the first synthetic life form. This breakthrough was only possible, though, by using sophisticated computer technology in order to digitally map out the complex coding inherent in tiny cells with simpler genomes such as the 582,970-base-pair M. genitalium.

Of course, Venter and his team would never have even started, much less succeeded, in their experiments, unless Watson and Crick had first unraveled the double helix structure to deoxyribonucleic acid, the basic building block to all life as we know it on this planet. Hence, even the most integrated of circuits must first be understood by unmasking its most fundamental of parts. This holds true whether one is compiling the *Oxford English Dictionary* (with its 26 letters from A to Z), developing the rudiments of computer processing (resting as it does on a binary number system), or making a wood fried pizza (dependent as it is on a preconceived list of ingredients). Thus, and contrary to popular misconceptions, the informational approach— whether in genetics or neuroscience—evolves out of reductionism not in juxtaposition with it and should not be conflated with vitalism or metaphysics.

As Venter explains, "DNA was the software of life, and if we changed that software, we changed the species, and thus the hardware of the cell. This is precisely the result that those yearning for evidence of some vitalistic force feared would come out of good reductionist science, of trying to break down life, and what it meant to be alive, into basic functions and simple components. Our experiments did not leave much room to support the views of the vitalists or of those who want to believe life depends on something more than a complex composite of chemical reactions."

Likewise, Christof Koch, who worked for years with Francis Crick on consciousness (Crick in 1994 dedicated his ultimate reductionist manifesto, *The Astonishing Hypothesis*, to Koch) and who was for many years a Professor of Biology at Cal Tech, hasn't abandoned reductionism simply because he strongly believes in Tononi's "Phi" and the way of "integrated information." To the contrary, Koch recently resigned from Cal Tech to take up his current position as Chief Scientific Officer of the Allen Institute for Brain Science in Seattle,

Washington, where among other projects he focuses on studying the neural correlates necessary for human consciousness. A cursory survey of the institute's sponsored publications should assuage any fears one may have that a Phi approach is incompatible with intertheoretic reductionism:

Huang JZ & Zeng H. (2013). Genetic approaches to neural circuits in the mouse, *Annual Review of Neuroscience* doi:10.1146/annurev-neuro-062012-170307

Krey JF, Pasca SP, Shcheglovitov A, Yazawa M, Schwemberger S, Rasmusson R, Dolmetsch RE. (2013). Timothy syndrome is associated with activity-dependent dendritic retraction in rodent and human neurons, *Nature Neuroscience* doi:10.1038/nn.3307

Stottmann RW, Donlin M, Hafner A, Bernard A, Sinclair DA, Beier DR. (2013). A mutation in Tubb2b, a human polymicrogyria gene, leads to lethality and abnormal cortical development in the mouse, *Human Molecular Genetics* doi:10.1093/hmg/ddt255

It should also be noted that information theory besides now being elemental in genomics and neuroscience is also championed by physicists as a fundamental way of understanding the quantum universe.

As Seth Lloyd explained in his now classic tome, *Programming the Universe*: "The universe is made of bits. Every molecule, atom, and elementary particle registers bits of information. Every interaction between those pieces of the universe processes that information by altering those bits. That is, the universe computes, and because the universe is governed by the laws of quantum mechanics, it computes in an intrinsically quantum mechanical fashion; its bits are quantum bits. The history of the universe is, in effect, a huge and ongoing quantum computation. The universe is a quantum computer."

If such is true, then what we are witnessing in differing scientific disciplines is how varying levels of computation (or information processing) evolve over time into sophisticated complex systems, ranging from the elements in the periodic table to mutating viruses and bacteria to self-conscious animals to artificial intelligence guided by digital software.

What this portends is a holistic way of understanding matter and mind not by appealing to mythic animism but rather by studying how intersecting bits of information cohere and create replicating forms of intelligent organisms. Or, to put it more precisely, reductionism and integration are not dueling alternatives, but rather complementary pathways that necessitate each other.

The implications of this underlying informational approach, however, are mind boggling to say the least.

In genomics, Craig Venter predicts that in the future we will be able to transfer a complete digital blueprint via biological teleportation: "When we read the genetic code by sequencing a genome, we are converting the physical code of DNA into a digital code that can be transformed into an electromagnetic wave that can be transmitted at the speed of light."

"The day is not far off when we will be able to send a robotically controlled genome-sequencing unit in a probe to other planets to read the DNA sequence of any alien microbe life that may be there, whether it is living or preserved."

Already Craig Venter has had his own genome sequenced and broadcast into space. This was only made possible because of the unique configuration of his DNA that could be translated into a digital record. Because Venter's genotype sequencing, as such, was substrate neutral it allowed for a binary informational upload.

If evolving forms of consciousness, like all biological phenotypes, has an informational blueprint of its own, then it too can be uploaded and transferred digitally provided that it is ultimately computational. In this scenario self-reflective awareness is ultimately but one aspect of integrated information processing. This indicates that there can and must be a wide spectrum of differing internal states that correlate with complex chunks of matter, be it biological or digital. This is what is meant by consciousness being substrate neutral.

As Koch elaborates, "We live in a universe where organized bits of matter give rise to consciousness. And with that, we can ultimately derive all sorts of interesting things: the answer to when a fetus or a baby first becomes conscious, whether a brain-injured patient is conscious, pathologies of consciousness such as schizophrenia, or consciousness in animals."

We may also be able in the future to do precisely what Ray Kurzweil and other futurists have long prophesized, which is to reverse engineer the human brain and then transfer that information algorithmically into electromagnetic waves with the possibility of reconstructing it into different mediums— mediums which are not as biologically brittle as the human body.

Nevertheless, Sir Roger Penrose argues that consciousness may be the result of quantum properties and as such cannot ad hoc be reduced computationally since there are non-algorithmic features inherent in the subatomic world. Others such as the Nobel Prize winning neurophysiologist, Sir John Eccles, believed that the mind (or soul) was something quite distinct from the brain that housed it.

As he stridently argued in *Evolution of the Brain, Creation of the Self*, "I maintain that the human mystery is incredibly demeaned by scientific reductionism, with its claim in promissory materialism to account eventually for all of the spiritual world in terms of patterns of neuronal activity. This belief must be classed as a superstition. . . . we have to recognize that we are spiritual beings with souls existing in a spiritual world as well as material beings with bodies and brains existing in a material world."

Yet, Eccles' dualism has tended to be ignored by the general scientific community as a non-starter since a "soul" theory by definition is metaphysical and not readily amenable to confirmation or refutation. As Patricia Smith Churchland argued in her recent book, *Touching a Nerve*, "Back to my wisdom tooth. Can the dualist match neuroscience's level of explanatory consilience regarding why procaine blocks pain? Not even close. A dualist could say, well, the procaine also acts on the soul. But how, even roughly, does that work? What does it do to the soul—especially if procaine is physical and the soul completely not physical? This dualist says

129

nothing at all about mechanism. Consider the contrast with the neuronal explanation, which is all about mechanism."

What is perhaps most exciting today about the scientific quest to understand consciousness is that so many avenues (even if contradictory) have opened up and been championed in various quarters, whereas in the past the subjective nature of human awareness was regarded as more or less a Skinnerian "black box."

Issac Asimov, not surprisingly given his prophetic track record with regard to most things scientific, captured the essence of informational theory when he analogized that selfhood was akin to a sand castle on a beach. Yes, it is made of tiny grains of sand, but the architecture cannot be reduced to just one bit since it is the totality of how those bits are compiled that makes all the difference. Alter those grains and you can construct a moat; alter them yet again and you can reproduce a tower or an underground tunnel. Similarly, the self is the result of a vast network of intersecting bits of matter, including neurons, synapses, dendrites, axons, and chemical electrical fluid, etc. Without those integrated neuronal switches human consciousness, (as we presently know it) doesn't manifest, just as the sand castle doesn't appear without its constituent parts being intact. However, if consciousness like its sand counterpart is--in terms of informational processing--substrate neutral then one could, if he or she so desires, reverse engineer its coordinates and complex intersections and reconstruct it anew in an entirely different medium, provided such reconstructions were digitally accurate.

We have already seen what this new brave new world of information processing has done (and will continue to do) in the world of genomics. Since we have already created synthetic cells based on DNA coding, one can only wonder if in the not so distant future whether scientists will be able to construct a truly synthetic self.

As Seth Lloyd concludes about the universe at large, "The primary consequence of the computational nature of the universe is that the universe naturally generates complex systems, such as life. Although the basic laws of physics are comparatively simple in form, they give rise, because they are computationally universal, to systems of enormous complexity."

I thoroughly enjoyed Andrew P. Smith's recent essay, "Consciousness So Simple, So Complex". I thought it was quite well written and demonstrated a deep engagement with Tononi's and Koch's Integrated Information Theory concerning consciousness. I particularly liked Smith's detailed distinction between unconscious and conscious processes and the necessity of the former for the latter. Writes Smith, "So while our conscious experience is indeed highly integrated, a great deal of that integration takes place unconsciously. Consciousness is often described as 'where it all comes together', but it has already come together to a considerable degree before we are conscious."

What most caught my attention, however, was Smith's invocation of what has been called in various intellectual quarters as the "Zombie" argument, which has parallels (even if not precisely) with Searle's Chinese Room argument.

As Smith explains, "One could imagine a zombie—a being behaviorally identical to a human being but lacking in consciousness—playing a musical instrument in exactly the same fashion as an ordinary human. Its brain would still have to unify the different sensory modalities—if it didn't, it couldn't perform the musical composition correctly—and we could certainly imagine something like Koch's view of integrated information enabling it to do so. But the zombie would still lack conscious experience of playing the instrument. Comparing the zombie to an actual human being, there is something missing, and that missing something is not obviously provided for by integrated information."

The Zombie dilemma, of course, raises the age-old philosophical issue of "other" minds. While I can be quite certain of my own subjective experience—the qualia of this or that occurrence—the same cannot be said about my perception of other beings. In other words, I may have great confidence in my own self-reflective awareness, but when it comes to "you" (the other) I am not so sure.

I have in a series of articles touched upon this conundrum (particularly in "Is My iPhone Conscious?" and "The Disneyland of Consciousness") since it is one that seems elemental to why such awareness may have evolved in the first place.

As humans we have a tendency to impute conscious intentionality upon a whole host of materials outside of ourselves. Depending on the time and the space, we may animate the sun, moon, and an array of planets and stars. Or, we may impute soul-like qualities to mountains, rivers, trees, and even certain precious stones or metals

Daniel Dennett, the distinguished professor of philosophy at Tufts University, has long argued that taking "intentional stances" allows us to develop predictive models of how others may behave. As the *Conscious Entities* website clearly explains,

"According to [Daniel Dennett] there is, in the final analysis, nothing fundamentally inexplicable about the way we attribute intentions and conscious feelings to people. We often attribute feelings or intentions metaphorically to non-human things, after all. We might say our car is a bit tired today, or that our pot plant is thirsty. At the end of the day, our attitude to other human beings is just a version--a much more sophisticated version--of the same strategy. Attributing intentions to human animals makes it much easier to work out what their behaviour is likely to be. It pays us, in short, to adopt the intentional stance when trying to understand human beings."

Taking such intentional stances, however, doesn't mean that the object we are currently animating (lightening, say) necessarily has that particular attribute. No, it is just that such psychic transferences provide us with a stratagem with which to act or react.

This naturally leads to the contentious issue of Zombie consciousness, since the very word conjures up all sorts of definitions ranging from the rudimentary, "a person who moves very slowly and is not aware of what is happening especially because of being very tired" (which sounds like my brother early on Saturday morning) to the specifically magical, "a will-less and speechless human in the West Indies capable only of automatic movement who is held to have died and been supernaturally reanimated."

What seems to be common among almost all Zombie definitions is that the person looks and acts a human being, but inside is soul-less or unconscious of what he or she is doing—which, in most horror movie versions, is when it is trying to eat another person, who is usually all too aware of what is happening!

Yet, this begs a much larger question: how do we really know that the Zombie is merely a mindless automaton? The obvious answer is that we don't. For instance, medical doctors have only recently come to realize that some coma patients who they mistakenly believed were in a completely vegetative, non-aware state, were in fact quite conscious but were unable to communicate such outwardly.

The case of Ron Houben is chilling case in point. As the *Daily Mail* in U.K. reported, "A car crash victim has spoken of the horror he endured for 23 years after he was misdiagnosed as being in a coma when he was conscious the whole time. Rom Houben, trapped in his paralysed body after a car crash, described his real-life nightmare as he screamed to doctors that he could hear them - but could make no sound. 'I screamed, but there was nothing to hear,' said Mr Houben, now 46, who doctors thought was in a persistent vegetative state. 'I dreamed myself away,' he added, tapping his tale out with the aid of a computer. Doctors used a range of coma tests before reluctantly concluding that his consciousness was 'extinct'. But three years ago, new hi-tech scans showed his brain was still functioning almost completely normally. Mr Houben described the moment as 'my second birth'. Therapy has since allowed him to tap out messages on a computer screen. Mr Houben said: 'All that time I just literally dreamed of a better life. Frustration is too small a word to describe what I felt.' His case has only just been revealed in a scientific paper released by the man who 'saved' him, top neurological expert Dr Steven Laureys. 'Medical advances caught up with him,' said Dr Laureys, who believes there may be many similar cases of false comas around the world. The disclosure will also renew the right-to-die debate over whether people in comas are truly unconscious."

The world of appearances can be a beguiling and deceiving arena indeed. We tend to make quick and ad hoc judgments when rushed so that if something walks like a duck, quacks like a duck, and looks like a duck, we are convinced it really is a duck, forgetting in the process that it may be nothing of the sort. We now live at a time where CGI effects are so convincing that it is becoming nearly impossible to distinguish a manufactured image from a real one.

We are fast approaching the time when the Turing test (in which a computational device will be able to trick us into believing that it is human) will be routinely passed and where the dividing line between artificial and human intelligence will most likely be forever blurred.

I raise this issue because I think the Zombie argument is a misleading one, since it assumes (wrongly, I suggest) that we know a priori that the creature in question is somehow devoid of what we ourselves possess. But how do we even know this, except by exterior observation, which is precisely how we adjudicate whether other humanoids have minds similar to our own. This is an epistemological cul du sac, since we at present don't have the ability to know what it is like to be a Zombie or even a bat, as Nagel cautioned decades ago. And if this is true then the Zombie hypothesis is a non-starter because we are stuck within the limits of our subjective universe. We simply cannot a priori deny the qualia of another, whether it is our lover, our dog, or a Zombie standing next to us at McDonald's ordering French Fries.

It appears we are trapped (at least temporarily) within a biological version of Plato's *Allegory of the Cave*, where we are bound to interpreting mere tracings and then inferring what such sketching may portend.

This it can be argued is the real hard problem of consciousness. As Sam Harris, notwithstanding his neuroscience background, admits, "The problem, however, is that no evidence for consciousness exists in the physical world. Physical events are simply mute as to whether it is 'like something' to be what they are. The only thing in this universe that attests to the existence of consciousness is consciousness itself; the only clue to subjectivity, as such, is subjectivity. Absolutely nothing about a brain, when surveyed as a physical system, suggests that it is a locus of experience. Were we not already brimming with consciousness ourselves, we would find no evidence of it in the physical universe—nor would we have any notion of the many experiential states that it gives rise to. The painfulness of pain, for instance, puts in an appearance only in consciousness. And no description of C-fibers or pain-avoiding behavior will bring the subjective reality into view."

John Searle tackles this dilemma using a linguistic pointer when he argues that any and all 3rd person descriptors cannot provide us with the internal view of our 1st person experiences.

Therefore, the complexity of any neural system though it may or may not be necessary for self-reflective consciousness cannot in itself (as a set of data points) provide us with what it is like to be a jelly fish, a gold ring, or an extraterrestrial, since subjective components cannot be captured objectively, try as we might. Harris elaborates on this when he argues,

"It is possible that some robots are conscious. If consciousness is the sort of thing that comes into being purely by virtue of information processing, then even our cellphones and coffeemakers may be conscious. But few of us imagine that there is 'something that it is like' to be even the most advanced computer. Whatever its relationship to information processing, consciousness is an internal reality that cannot necessarily be appreciated from the outside and need not be associated with behavior or responsiveness to stimuli. If you doubt this, you must read *The Diving Bell and the Butterfly*, Jean Dominique-Bauby's astonishing and heartbreaking account of his own "locked-in syndrome"—which he dictated by signing to a nurse with his left eyelid—and then try to imagine what his predicament would have been if even this degree of motor control had been denied him."

In this way we seem to be stuck in a behaviorist's locked vault. All we can surmise about anyone else is our own intentional stances projected outwardly that we confirm or disconfirm by our increasingly sophisticated observations, which are exponentially aided by our technological prowess. Yet, even here we are still caught within a neurological storefront that can only provide us with a menu of possible entry states but which cannot give us the taste of that specific internal realm. We are, in sum, onlookers to the subjective realm. Or, to bastardize a famous quote from Rudyard Kipling, "subjective is subjective and objective is objective and never the twain shall meet."

It is for this reason that humans have had a long history of mistaking something inanimate as animate and vice versus. Disneyland is full of animatronic characters (Zombies?) that have befooled many a visitor into thinking that they were real humans putting on a performance. Just as we can mistakenly believe that a comatose person is vegetative, we can also quite innocently presume that the animatronic in *Great Moments with Mr. Lincoln* on Main Street in Disneyland is a normal human being.

Therefore, the Zombie argument doesn't actually illuminate the hard problem of consciousness but only obfuscates the knotty issue of what it is like to be something other than what we ourselves our.

Andrew Smith illuminates on this when he writes, "In fact, much of our actual behavior is not that far removed from Zombieland. When an accomplished professional musician performs, most of what the brain does is carried out unconsciously. As has been recognized for decades, this is one of the major distinctions between an expert and a novice, with regard to any form of behavior. The novice initially has to be conscious of nearly every detail of the performance, whereas for the expert, someone who has practiced the behavior extensively in the past, the performance is mostly automatic. This is very clear evidence that a vast amount of integration of information in the brain is carried out unconsciously."

But perhaps there is no hard problem in consciousness as posited by David Chalmers and others and we have only made it hard or harder (as Dennett has suggested in his recent book, *Intuition Pumps and Other Tools for Thinking*) by claiming that it is.

In this contrarian view, the way to understand consciousness is to assume (metaphorically speaking) that we are all Zombies and that what we take to be so unique in our first person narratives is, on closer inspection, not as special or resistant to outside comprehension as we might presume.

Daniel Dennett has long championed the view that consciousness can be explained provided that we recalibrate how we view our own notion of what it means to have a self. As Dennett argues, "It's [consciousness] astonishingly wonderful but it is not a miracle and it isn't magic. It's a bunch of tricks and I like the comparison with magic because stage magic of course is not magic, magic. It's [a] bunch of tricks and consciousness is a bunch of tricks in the brain and we're learning what those tricks are and how they fit together and why it seems to be so much more than that bunch of tricks. Now, for a lot people the very suggestion that, that might be so is offensive or repugnant. They really don't like that idea and they view it as in a sort of an assault on their dignity or their specialness and I think that's a prime mistake."

One way to tackle this divided take on consciousness (whether a soft or hard problem) is to follow Gerald Edelman's lead and define awareness in a two-fold way: First nature, or primary awareness is of the present moment and the immediate past and can be understood as sensory consciousness or sentience itself; Second nature, or higher order awareness is when we can deeply reflect upon past, present, or future actions and have the ability to be cognizant of our self-consciousness—awareness of awareness, so to speak.

This can also be parsed as first nature being "association" (in the moment and in tune with prevalent proceedings) or "zoning in"; and "dissociation" (out of sync with what is occurring and ruminating about past or future activities) or "zoning out."

Analogously, primary consciousness is an engaged moment whereas secondary consciousness is disengaged to the point where one is somewhere else (mentally speaking).

Edelman makes these distinctions since he feels that most if not all animals have some sense of primary awareness or sentience whereas secondary or higher order awareness may be only the lot of those with more highly ordered and complex brain structures.

Why evolution would have evolved such a sophisticated form of awareness isn't as intractable as it may at first seem. Again, Andrew P. Smith touches upon this when he asks the question of why consciousness would have evolved when apparently an unconscious Zombie could have the same functional advantages.

Putting aside the ontological question of what Zombies may or may not experience subjectively, I think it is fairly obvious what evolutionary benefit higher order consciousness conveys upon those who have it. Any organism that can "virtually simulate" varying options within itself before outsourcing them in a real, empirical world has a tremendous advantage over creatures who lack such a simulacrum Rolodex.

Think about what your consciousness does most of the time, particularly during an uneventful and tedious lecture on consciousness itself. It spaces out. We tend to dissociate and ruminate about all sorts of things, from fantasizing about this or that person or imagining what we are going to do on vacation or perhaps if the lecture doesn't go too long about what we are going to order at Veggie Grille. These mind wanderings allow us to conjure up all sorts of real and unreal possibilities and as such allow us the opportunity to play out different end game scenarios without ending up injured or dead. Those who lack such an internal theatre don't have the ability to "rehearse" and thus are forced to act in a real and dangerous world almost immediately.

There is, undoubtedly, a drawback to have such a conjuring mind since it can (and often does) capture us in an admixture of fantastic phantasms which have no direct correlation to the eat or be eaten world in which we live.

We can also be subject to some truly horrifying delusions, such that it becomes nearly impossible to differentiate the exterior world from our interior machinations.

This became clearer to me this past semester at Mt. San Antonio College when I became well acquainted with an older student in my Introduction to Philosophy class. He introduced himself early in the semester after I had given a lecture on consciousness as a virtual simulator. He couldn't look me in the eye and explained that he hadn't left his house for nearly 20 years. He had been diagnosed with schizophrenia and had an inordinate fear of people. His doctor had advised him to go back to school and my class was the first one he was taking. He was mesmerized by the idea of the virtual simulator since he felt that it explained his situation to a tee.

He then proceeded to write a long narrative in which he described in excruciating detail what a typical day in his was like. Here is but one small excerpt:

"Now that I look back, I remember that I did experience mild panic attacks beginning at the tender age of nine years old. I would sleep in a sitting position because I thought I was going to vomit my intestines if I slept lying down. I slept in this position for months. Now that I look back at this moment, I realize that this was just the beginning of the nightmare that lay ahead. You see, once a stubborn notion enters my mind, I cannot get rid of it. It completely takes over my mind and body. I guess I am so screwed up in my head that when a notion enters my mind, I get sick for days. My body gets overwhelmed with fear. I also begin to tremble. My head begins to hurt and my stomach begins to turn. I sometimes even suffer from diarrhea. The sickness lets me know that a negative notion has entered my mind. If the notion was caused by wearing a new shirt, I either don't ever wear the shirt again, or I decide to put up with the sickness every time I wear it. I have been asking myself the same question for twenty years, "How can something as harmless as a shirt cause so much mental and physical pain?" I know the shirt is not to blame for my defective mind, but I would still love to know the scientific explanation from beginning to end. I always end up depriving myself of many simple pleasures in life because I have associated them with pain. Half of the things I own I don't use because a negative notion in my head developed that induces fear

141

every time I try to use these items. For example, I bought an expensive stereo a few years ago and I haven't touched it since then. I am afraid that my head is going to explode if I listen to the stereo (that is the crippling belief that exists in my mind). Every time I try to use these items, a rush of fear takes over me and I begin to perspire. My stomach begins to turn and I immediately begin to get sick. If I am ever in a rare happy mood, all I have to do is reach out for one of these items and fear and sickness will wreak havoc on my body and mind."

As I got to know this student (we would invariably talk at length after our class was over) I realized how brilliant he was. Indeed, he was by far the brightest student I had taught in years, since he seemed to grasp almost any complex subject immediately, despite not having any academic training per se.

What truly surprised me, though, was that once he understood the virtual simulator hypothesis it noticeably calmed him down because the model helped him better understand his own behavior and how he was fueling (even if unconsciously) his own obsessive behavior and getting trapped within it. Over the course of the term, he even began to make prolonged eye contact and to develop a sense of humor about his predicament. This student also claimed to have improved more in the past few weeks than he had in ten years of seeing his psychiatrist. It may well be that if we are convinced in the rightness of a theory and how it applies to our own situation, it can liberate us to some degree from our own guilt and our own consternations. For example, I have noticed that a medical condition can dramatically improve if I receive a proper diagnosis that I can firmly believe is accurate. Clearly, the placebo effect is a powerful one across all fields including those suffering from certain mental illnesses.

Higher order consciousness may provide a tremendous advantage for us to live within our heads before acting out our varying plans, but it can also boomerang against us since we have a tendency to conflate our dreams with the world around. To the degree that our internal models help us navigate and survive it is a tool of incalculable power, but to the degree that it binds us into mistakenly believing (without sufficient evidence) our own imaginings we can become prisoners within its hallucinatory walls.

It can be argued that we are all delusional to some extent since many of our delusions (religious or otherwise) allow us to buffer ourselves from the stark reality that the planet we find ourselves foraging about is a death machine where no one gets out alive. Consciousness as a virtual simulator (Edelman's 2nd nature) may have evolved not only to help us with developing ways to strategize but also to distract us from our precarious predicament. As my student once quipped to me, "Too much reality and we become catatonic and too much fantasy and we become schizophrenic."

Perhaps most importantly, certain aspects of the virtual simulator hypothesis have been tested and are garnering impressive results. *New Scientist* recently reported that thinking about a certain activity, such as juggling skills, before doing it could significantly improve one's ability later on:

"Sook-Lei Liew and her colleagues from the National Institute of Neurological Disorders and Stroke in Bethesda, Maryland, asked eight adults to watch a circle on a screen while an fMRI machine scanned their brain. When the circle turned into a triangle, the volunteers moved their fingers. This movement caused activity in their premotor cortex and supplementary motor cortex – brain areas involved in imagining moving and actually moving – which in turn raised a bar on the screen. The more synchronised the brain activity, the higher the bar went. More synchronisation has previously been linked with better performance in movement tasks. The researchers then asked the volunteers to imagine performing a complicated action – whatever they liked, as long as it increased the height of the bar. This enabled them to develop a way of improving coordination between the brain regions using thought alone. After an hour of mental practice, participants were 10 per cent faster at a manual task. Those who showed the greatest increase in speed also showed the greatest enhancement in synchronisation."

Interestingly, some neuropsychologists believe that our sense of a self developed after (not before) we evolved ways to envisage (via intentional stances) why others behaved the way they do. In other words, *we became adept at projecting prior to becoming adept at reflecting.* Ironically, this inverse reasoning implies that Zombie consciousness precedes self-awareness, since the former (unconscious processing) is a forerunner of the latter. If our nocturnal sojourns are an indication of our ancestral past, then it might be correct to surmise that we learned to dream before we could scheme.

Smith rightly touches upon the importance of unconscious processing and integration as a necessary prelude to conscious self-awareness and suggests that any theory of consciousness should take into consideration that what is unconscious to us as humans may be "conscious" to animals devoid of higher order awareness

It is certainly true that whenever we learn a new skill, such as surfing, we have to be keenly aware of almost everything we do at first: how to paddle, how to sit on a board, how to catch a wave, how to stand up quickly as the wave crashes, how to bottom turn, and so on. But once we have mastered such a set of skills they become automatic reflexes and we act more instinctively. We no longer need to pay such close attention to them, as they have become unconscious reflexes, so to say. Tom Curren, a three-time world champion surfer, once explained that his very best surfing was when he was mostly mindless and instead of being conscious of each and every turn he simply rode the wave and responded accordingly. But this can only happen after many hours of concentrated attention.

In this way, one could argue that the Zombie argument of consciousness should be placed on its head (and the bad pun here is purely intentional), since we only become Zombie like after (not before) being fully aware.

In this context, therefore, Zombie consciousness doesn't necessary indicate that a person or object is mindless at all, but rather that they might have been so mindful at one stage that they later earned the luxury of tuning and zoning out whenever possible. This may explain why all those people driving cars after work in downtown Los Angeles look like Zombies from the "Night of the Living Dead."

About the Authors

Andrea Diem-Lane is a Professor of Philosophy at Mt. San Antonio College. She received her Ph.D. and M.A. in Religious Studies from the University of California, Santa Barbara, where she did her doctoral studies under Professor Ninian Smart. Professor Diem received a B.A. in Psychology with an emphasis on Brain Research from the University of California, San Diego, where she did pioneering visual cortex research under the tutelage of Dr. V.S. Ramachandran. Dr. Diem is the author of several books including an interactive textbook on religion entitled *How Scholars Study the Sacred* and an interactive book on the famous Einstein-Bohr debate over the implications of quantum theory entitled *Spooky Physics*. Her most recent book is *Darwin's DNA: An Introduction to Evolutionary Philosophy*. Andrea and David are the parents of two boys, Shaun-Michael and Kelly-Joseph.

David Christopher Lane is a Professor of Philosophy at Mt. San Antonio College and an Adjunct Lecturer in Science and Religion at California State University, Long Beach. He received his Ph.D. in the Sociology of Knowledge from the University of California, San Diego, where he was also a recipient of a Regents Fellowship. He has taught previously at Warren College at UCSD, the University of London, and the California School of Professional Psychology. He has given invited lectures at various universities, including the London School of Economics. He is the author of a number of published books such as The *Making of a Spiritual Movement: The Untold Story of Paul Twitchell and Eckankar*; The *Radhasoami Tradition: A Critical History of Guru Succession*; *Exposing Cults: When the Skeptical Mind Confronts the Mystical*; and *The Unknowing Sage: The Life and Work of Baba Faqir Chand*, among others. David and Andrea are currently working on a large book and film project entitled, *The Library of Consciousness: Exploring a Multiverse of Selves*.

Made in the USA
San Bernardino, CA
04 September 2014